Farm Machinery & Power

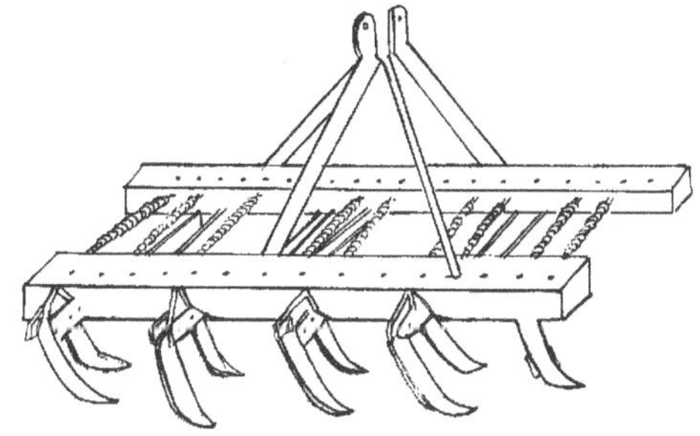

NIPA® GENX ELECTRONIC RESOURCES & SOLUTIONS P. LTD.
New Delhi-110 034

www.nipaers.com

Browse, Search, Read & Buy...

Print Books eChapters

eBooks Publishers

Forthcoming Subject Catalogues

New Titles

Online Resources on:

- Current Affairs
- Reasoning, Logic & Aptitude
- Competitive Examinations
- English Language Lab
- Writing & Pronunciation Tools
- Personality Development
- Online Programmes for Professionals
- Interview Preparation

Pay Using

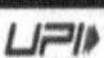

CC Avenue

Farm Machinery & Power

Glossary

Ashok G. Powar (Dean)
Faculty of Agricultural Engineering

&

Vijay V. Aware (Associate Professor)
Department of Farm Machinery and Power

College of Agricultural Engineering and Technology
Dr. Balasaheb Sawant Konkan Krishi Vidyapeeth
Dapoli, District Ratnagiri - 415 712
Maharashtra

NIPA® GENX ELECTRONIC RESOURCES & SOLUTIONS P. LTD.
New Delhi-110 034

NIPA® GENX ELECTRONIC
RESOURCES & SOLUTIONS P. LTD.

101,103, Vikas Surya Plaza, CU Block
L.S.C. Market, Pitam Pura, New Delhi-110 034
Ph : +91 11 27341616, 27341717, 27341718
E-mail:newindiapublishingagency@gmail.com
Web: www.nipabooks.com

For customer assistance, please contact
Phone: + 91-11-27 34 17 17
Fax: + 91-11- 27 34 16 16
E-Mail: feedbacks@nipabooks.com

© 2023, Publisher

ISBN: 978-81-19103-63-8

All rights reserved. No part of this publication may be reproduced, stored in a retrieval syste or transmitted in any form or by any means, including electronic, mechanical, photocopyi recording or otherwise without the prior written permission of the publisher or the copyrig holder.

This book contains information obtained from authentic and highly reliable sources. Reasonat efforts have been made to publish reliable data and information, but the author/s, editor/s a publisher cannot assume responsibility for the validity, accuracy or completeness of all materi or information published herein or the consequences of their use. The work is published wi the understanding that the publisher and author/s are not attempting to render any professior services. The author/s, editor/s and publisher have attempted to trace and acknowledge t copyright holders of all material reproduced in this publication and apologize to copyrig holders if permission and/or acknowledgements to publish in this form have not been taken. any copyrighted material has not been acknowledged, please write to us and let us know so th we may rectify the error, in subsequent reprints.

Trademark Notice: NIPA®, the NIPA® logos and their presentations (the way they are writte presented) in this book are the trademarks of the publisher and hence may not be used withc written permission, if copied or used without authorization, the infringer will be prosecuted per law.

NIPA® also publishes books in a variety of electronic formats. Some content that appears print may not be available in electronic books, and vice versa.

Composed and Designed by NIPA®.

Preface

Farm mechanization is the only way out for all the issues plaguing agricultural operations. Agriculturists should take to farm mechanization technique to enhance both productivity and profitability. A huge investment on agricultural labour and a sharp fall in labour efficiency posed a serious threat to crop management techniques. The mission of many government and private organizations is to make agricultural operations less drudgerious by introducing various tools, implements and machineries and hence to increase the farm mechanization index. These efforts involve making each and every term involved technically sound. Prerequisites for this is explaining the terms in understandable in scientific language and in order to lay the strong foundation of basic terms in farm machinery and power, the book **"Farm Machinery and Power: Glossary"** is compiled so that the reader should be able to understand and comprehend it better. This book will be very basic requirement for the students, academicians and persons engaged in farm mechanization. Those who enter and work in this field, go with some basic knowledge to implement it with confidence and efficiency.

There may be several mistakes and omissions. We earnestly request the readers to kindly intimate those with their views to improve in future. It will be a matter of great satisfaction to us if our efforts are found to be some help to the students, scientists and professionals in the field of farm mechanization.

The authors record the help of New India Publishing Agency, New Delhi, who have brought the manuscript to bring it to light. We owe to all who have helped us in various ways.

August 2006
Dapoli

Ashok G. Powar
Vijay V. Aware

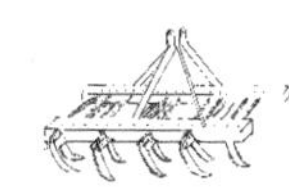

Contents

Preface *v*

A 1-18

Abnormal combustion – Axle load

(Comprising of 9 figures)

B 19-36

Babbitt – Bypass

(Comprising of 12 figures)

C 37-72

Cable conveyor – Cylinder type chaff cutter

(Comprising of 22 figures)

D 73-89

Dah – Dynamometer

(Comprising of 14 figures)

E 91-108

Ear plug – Eyeleting

(Comprising of 3 figures)

F 109-140

Fabrication – Fusion piercing drill

(Comprising of 15 figures)

G 141-158

Gadget – Gyroscopic

(Comprising of 15 figures)

H 159-176

Half round file – Hythane

(Comprising of 9 figures)

I 177-186

I beam – Isothermal process

(Comprising of 3 figures)

J 187-189

Jack, electric – Junker's engine

K 191-194

Kelvin scale – Knurling

(Comprising of 1 figure)

L 195-202

Labeled – Lumber

(Comprising of 3 figures)

M 203-218

Mach number – Multistage expansion with reheating

(Comprising of 3 figures)

N 219-224

Natural gas – Nut

O 225-233

Oak – Oxygen (O_2)

(Comprising of 3 figures)

P 235-264

Packing – Pyrometer

(Comprising of 6 figures)

Q 265-266
Quadrangle – Quoins

R.. 267-286
Rabbet – Runway
(Comprising of 9 figures)

S 287-327
SAE – Synthetic
(Comprising of 9 figures)

T.. 329-344
T bolt – Two-stroke cycle
(Comprising of 6 figures)

U 345-347
U clamp – Uranium

V 349-356
V belt – Volume fraction
(Comprising of 1 figure)

W 357-363
Walking machine – Wrist pin
(Comprising of 1 figure)

XYZ 365-366
Xengine – Zinc

Bibliography 367

Abnormal combustion: Combustion that does not proceed in the normal way, in which knock, pre-ignition, run-on or surface ignition occurs. *Refer* knock.

Abrasion: It is the surface discontinuity brought to any solid by the process of removal of surface material through friction of another solid, liquid or gas or combined thereof.

Absolute or specific humidity: It is the mass of water vapour present in a unit mass of dry air. It is also called as humidity ratio.

Absolute instrument: It is an instrument, which measures a quantity (such as pressure, temperature etc.) in absolute units by means of simple physical measurements on the instrument.

Absolute pressure: It is the actual pressure at a given position and it is measured relative to absolute vacuum (i.e., absolute zero pressure). Absolute pressure = pressure shown on the pressure gauge + atmospheric pressure. At sea level, atmospheric pressure is 14.7 psi (1.033 kg/cm^2).

Absolute specific gravity: Ratio of the weight of a given volume of a substance to the weight of an equal volume of water in a vacuum at a given temperature.

Absolute strength: It is the actual breaking strength of a body.

Absolute temperature: Temperature scale with absolute zero as the zero of the scale. Absolute zero is referred to Rankine or R, and in metric as Kelvin or K. In standard, the absolute temperature R is the temperature in °F plus 460 (R = ^{0}F + 460), or in metric, it is the temperature in °C plus 273.15 (K = ^{0}C+ 273.15). The relation in ^{0}C and ^{0}F is, ^{0}C = [(^{0}F – 32)x 5/9]

Absolute viscosity: The tangential force per unit area of two parallel plates at unit distance apart when the space between them is filled with unit velocity in its own plane relative to the other. Also known as coefficient of viscosity.

Absolute zero: The temperature (–273.15 °K, or 0° C), at which molecular motion vanishes and a body would have no heat energy.

Absorption: Process by which a substance is drawn or included into the another substance.

Abstract design: Design, which is based on geometric lines and shapes.

Abrasivity: The property of particles of bulk materials, when in motion, to wear away the surface they are in contact with.

AC generator (or alternator): An electric device produces an electric current that reverses direction many times per second. Also called a synchronous generator.

Accelerated motion: Motion in which velocity is not constant. This term is used in reference to both increased and decreased velocity. When velocity is decreased, it is known as negative accelerated motion.

Accelerating jet: The carburetor jet through which the accelerating charge (fuel) is injected into the incoming charge (air). *See* Fig.1A.

Accelerating pump: A plunger type pump attached to the carburetor for the purpose of increasing the richness of the mixture in order to give a quick pick-up under load. *See* Fig.1A

Acceleration: The rate of change of velocity of a moving body.

Acceleration analysis: A mathematical technique (often done graphically), by which accelerations of parts of a mechanism are determined.

Accelerator: A mechanical device used for regulating the amount of charge, which is fed to the engine. It is usually foot operated.

Accelerator linkage: The linkage connecting the accelerator pedal of an automotive vehicle to the carburetor throttle valve or fuel injection control.

Accelerator pedal: A pedal that operates the carburetor throttle valve or fuel injection control of an automotive vehicle.

Accelerometer: An instrument for recording, measuring, or indicating acceleration.

Accelerometry: The quantitative determination of acceleration and deceleration in the entire human body or part of body in the performance of a task.

Accumulator: A cylinder into which water is forced in order to furnish the motive power in hydraulic machines of various kinds.

Accuracy: Closeness of the instrument output value to the true value. In practice, it is specified as the percentage deviation or inaccuracy of the instrument from the true value. The accuracy of the instrument is expressed as 'percentage of true value'.

$$\text{Percentage of true value} = \frac{(\text{measured value - true value})}{\text{True value}} \times 100$$

Acetone: An inflammable liquid (CH_3COCH_3) with a bitter taste, obtained by the destructive distillation of certain wood, acetates, and various organic compounds.

Acid: 1. A substance, which produces hydrogen ions when dissolved in water. 2. A hydrogen compound whose aqueous solution contains hydrogen ions and no other positive ions.

Acid rain: The falling rain (or snow), which is acidic in nature (pH value less than 5.6) due to its gaseous pollutants, such as oxides of sulfur and nitrogen.

Ackerman steering gear: Differential gear or linkage that turns the two steered road wheels of self-propelled vehicle so that all wheels roll in circles with a common centre.

Activity chart: A tabular presentation of a series of operations of a process plotted against a time scale.

Actual capacity (farm machinery)**:** It is the area of field covered by a machine per unit time. It is expressed in hectare per day, acre per hour, etc.

Actual power: It is the power available at the output engine shaft. It is known as actual horsepower.

Acute: Sharp at the end; ending in a sharp point; pointed.

Acute angle: Angle less than 90°.

Addendum: It is the radial distance of a tooth from the pitch circle to the top of the tooth. *Refer* Fig.9G.

Addendum circle: It is the circle drawn through the top of the teeth and is concentric with the pitch circle. *Refer* Fig.9G.

Adhesion: 1. It is the property, which enables one surface to adhere or stick to another. 2. It is the molecular attraction between the molecules of different substances (e.g., water & glass).

Adhered soil cone: Masses of soil, which adhere on soil working surfaces and act as a part of the tool. Soil bodies may be stationary or in a relatively slow motion.

Adiabatic: The word adiabatic has come from the Greek word *adiabatos*, which means not to be passed.

Adiabatic compression: A reduction in volume of a substance without heat flow, in or out.

Adiabatic cooling: A process in which the temperature of a system is reduced without any heat being exchanged between the system and it's surrounding.

Adiabatic curve: It is a curve, representing the pressure and volume variations of a fluid, being expanded or compressed adiabatically.

Adiabatic expansion: Increase in volume without heat flow, in or out.

Adiabatic process: Any thermodynamic process, which takes place in a system without the exchange of heat with the surroundings. It is also called as 'isentropic process'.

Adjustable spray nozzle: Hydraulic spray nozzle, designed so that the shape of the spray may be altered without changing the components.

Admission port: The passage by which charge (air - fuel mixture) enters the engine.

Adsorption: The collection or accumulation of one substance on the surface of another substance.

Advance: To being forward; to move towards front; to make to go on.

Advanced ignition: Admitting the spark earlier in the cylinder of internal combustion engine, so that it will ignite the charge earlier. As the speed of engine increases, it is necessary to have an earlier spark.

Aerosols: These are the droplets having the 'volume mean diameter' below 50 μm.

After Bottom Dead Center or ABDC**:** It is the position of the piston in the cylinder bore after its lowest point in the stroke (Bottom Dead Centre, BDC). ABDC is measured in degrees of crankshaft rotation after BDC. *See* Fig.2A.

After Top Dead Center or ATDC: It is the position of the piston in the cylinder bore after its highest point in the stroke (Top Dead Centre, TDC). ATDC is measured in degrees of crankshaft rotation after TDC. *See* Fig.2A.

Aggregates: Conglomerations of primary soil particles produced by natural processes.

Agitation: An operation, which produces and maintains uniform spray mixture in the tank, and in the case of dusts of granules to facilitate their flow from the hopper.

Agitator: A device used to agitate.

Agitator feed mechanism: A feed or distributing mechanism or both consisting of an agitator which cause a substance (fertilizer or seed) to flow evenly through an adjustable aperture. *See* Fig.3A.

Agitator wheel drill: A seed drill, the metering mechanism of which consists of agitator wheels. The seed rate is controlled mainly by varying the discharge apertures in the hopper.

Agricultural Engineering: A branch of engineering dealing with physical problems involved in land preparation for crop production, their processing and storage.

Agricultural tractor: A self propelled wheeled vehicle having two axles or a track laying machine, more particularly designed to pull, push, carry and operate equipment and machines used for agricultural work.

Agricultural trailer: A transport vehicle used for carrying agriculture produce or articles and which according to its design is suited and intended for being coupled to an agricultural tractor.

Agriculture: It is the science, art, and business of cultivating soil, producing crops, and raising livestock.

Agro meteorology: A science dealing with climatologically conditions, which have direct relation or relevance to agriculture.

Agro-climatic regions: The grouping of different physical areas within the country in to broadly homogeneous zones based on climatic and adaphic factors.

Agro-ecology: The study of the relation of agricultural crops and environment.

Agro-forestry: A self sustaining and land management system, which combines production of agricultural crops, tree crops and live stock simultaneously or sequentially, on the same unit of land.

Agronomy: A specialized branch of agriculture science concerned with the theory and practice of field crop production and soil management.

Air assisted sprayer: A sprayer, in which, chemical in the liquid form is atomized in the form of small droplets and carried to the target by a stream or flow of air.

Air chamber: A vessel installed on piston pumps to minimize the pulsating discharge of liquid pumped. The chamber contains air under pressure. The air acts as a cushion to lessen the fluctuation of the liquid flow between the suction and delivery strokes of the piston. Also called as 'air vessel'.

Air cleaner: A device, which cleans air before it reaches an engine cylinder by filtering and removing dust, moisture and foreign matters.

Air compressor: An appliance for raising the pressure of air above that of the atmosphere.

Air cooling: The system of cooling an internal combustion engine cylinder by directing air against its surface having fins.

Air cooled engine: An engine cooled directly by a stream of air without the interposition of a liquid medium.

Air cycle: A refrigeration cycle, in which the working fluid remains as a gas throughout the cycle rather than being condensed to a liquid.

Air deflector: A device to alter the direction of the airflow.

Air engine: Engine in which, air is first compressed in the cold part of the cylinder, then transferred to the hot part of the cylinder where it is quickly heated and expanded. The power is generated due to expansion of air.

Air fuel mixture: In a carburetor gasoline engine, air and fuel is mixed in the appropriate ratio in the carburetor and subsequently fed into the combustion chamber.

Air-fuel ratio: The proportions, by weight, of air and fuel produced by carburetor or injector and supplied for combustion.

Air gauge: An instrument for ascertaining air pressure.

Air inlet valve: A valve, through which air is admitted in the cylinder of four-stroke diesel engine. *See* Fig.4A.

Air lock: A stoppage in a fuel or gas line caused by the bubble of air being pocketed somewhere along the line.

Air pollution: Any contamination of the air that is harmful to human, animal or plant.

Air valve: A valve controlling the passage of air, which is used in gas producers, blast furnaces, etc.

Alcohol: A colorless organic compounds, containing a hydroxyl (OH) group. The simplest alcohol is methanol (CH_3OH).

Algorithm: A process or procedure used to solve a problem. It is synonymous with 'programs' or 'software' in a computer system.

Align: To bring parts into proper position in relation to one another.

Alignment of mower: The cutter bar is set at about 88^0 to the direction of motion i.e. inward lead of 2^0 is given to it, in order to overcome the back pushing action of the crops. Generally 2cm lead per metre length of cutter bar is kept.

Alternating current (AC): An electric current that flows from positive to negative and from negative to positive in the same conductor.

Alternate plough: A plough having an even number of bodies turning to right and left respectively and mounted side-by-side, which can be raised or lowered separately, turning about transverse axis. *See* Fig.5A.

Alternative fuel: A fuel alternative to gasoline or diesel fuel that is not produced in a conventional way from crude oil, for example CNG, LPG, LNG, ethanol, methanol and hydrogen.

Ambient temperature: The temperature of the surrounding atmospheric air.

Ampere (Amp or A): 1. Ampere is that constant current which, if maintained in two straight parallel conductors of infinite length, of negligible cross section, and placed 1 metre apart in vacuum, would produce between these conductors a force equal to 2×10^{-7} Newton per metre of length (*Comitë International des Poids et Mesures*). 2. Unit of current flow. One Amp equals one Coulomb of charge passing by a point per second.

Analog signal: An electrical signal that varies in voltage within a given parameter.

Anchoring: A tillage operation used to partially bury and thereby preventing movement of foreign materials such as plant residue or paper mulches.

Anaemometer: An instrument for measuring the velocity of an air current.

Angle of approach (track tractor): Angle between ground and the section of track between the front bogie wheel and the front idler.

Angle of departure (track tractor): Angle between ground and the section of track between rear bogie wheel and the rear idler.

Angle of incidence: Angle at which a ray of light (or radiation) strikes a surface. It is measured between the incoming ray and a perpendicular to the surface at a point of incidence. *See* Fig.6A.

Angle of internal friction: The angle whose tangent is the coefficient of internal friction between particles. The limit of this angle is a right angle

Angle of nip: Angle between the roll faces at which a particle or grain will be just hold.

Angle of pitch: Rotation around a horizontal axis passing through coupling point and perpendicular to the longitudinal plane of symmetry of the vehicle. *See* Fig.7A.

Angle of reflection: Angle at which a reflected ray of light leaves a reflecting surface. It is measured between the outgoing ray and a perpendicular to the surface at a point of incidence. *See* Fig.6A.

Angle of refraction: Angle at which a refracted ray of light leaves the interface at which the refraction occurs. It is measured between the direction of the refracted ray and a perpendicular to the interface at the point of refraction. *See* Fig.6A.

Angle of repose: When a granular material is allowed to flow freely from a point into a pile, the angle, which the side of the pile makes with the horizontal plane, is called the angle of repose.

Angle of roll: Rotation around a horizontal axis passing through the coupling point and located in the plane of symmetry of the vehicle. *See* Fig.8A.

Angle of rupture: The angle at which any material, as iron or wooden beams, will break by application of load.

Angle of slide: It is an angle from the horizontal at which loose or fragmented solid materials starts to slide.

Angle of yaw: Rotation around a vertical axis passing through the coupling point. *See* Fig.9A.

Angular advance: An amount by which the angle between the crank of a steam engine and the virtual crank radius of the eccentric exceeds a right angle. Also known as angle of advance or angular lead.

Angstrom: An unit of length used in the measurement of the wavelength of short electromagnetic radiations, like X-rays. It is named for A.J. Angstrom (1814-1874), a Swedish physicist. $A^o = 10^{-8}$ cm.

Anhydrous ammonia applicator: Machine to apply ammonia in gaseous form at specified pressure.

Animal drawn Implement: An implement operated by animal(s).

Animal drawn vehicle: A loading platform fitted with at least two wheels and pulled by animals.

Anion: A negatively charged ion.

Anode: An electrode at which oxidation (a loss of electrons) takes place. Anode is electrically negative.

ANSI: American National Standards Institute.

Anti drip device: A device, which normally forms a part of or fitted within the nozzle, to instantly prevent any further flow or dripping from the nozzle after spray boom has been shut-off.

Anvil: 1. A component of deflection nozzle, which deflects the spray liquid after the emission from the nozzle orifice. 2. A steel iron block upon which forging is done.

Apparent velocity of air flow: The rate of air flow determined by dividing the quantity of air flow by the cross-sectional area.

Application rate (pesticide/insecticide)**:** The quality of spray mixture, dust or granules distributed by an appliance per hectare.

Archimedes' principle: The net upward or buoyant force, equal in magnitude to the weight of the displaced fluid, acts upon a body either partly or wholly submerged in a fluid at rest under the influence of gravity. Named for Archimedes, a Greek mathematician who discovered the principle.

Arrester: Any mechanical contrivance or device used to stop or slow up motion.

Aromatics: Chemical compounds added to natural gas in order to impart odour.

ASME code: The American Society of Mechanical Engineers' Boiler and Pressure Vessel Code.

Aspirator: A component used for cleaning the seeds by drawing the air through the seed.

Aspirating leg: A vertical duct in which the upward movement of a controlled air blast remove lighter grains and impurities from vertically falling grain.

Aspiration: It is the process of cleaning by air blast and separating the foreign material, which is substantially lower in specific gravity than the product to be cleaned.

Assemble: To connect the separate parts of an engine/ a machine.

Assembling: Connecting together the parts of an engine or a machine. The combination of the various parts forms an intricate article.

Assembly drawing: An engineering drawing depicting a complete unit having detail drawings of all parts.

Atmosphere: The gaseous envelope of air surrounding the earth, which contains a complex system of gases and suspended particles being derived from the earth by way of chemical and biochemical reactions.

Atmospheric pressure: The absolute pressure above a perfect vacuum at any geographic location or temperature. The atmospheric pressure at sea level is 76 cm of Hg or 14.74 psi or 1.0333 kg/cm^2 at 59 ^oF. Any change in altitude, temperature or movement of atmospheric air masses alters this figure. (1 atm=1bar=76cm of hg=1.0333 kg/ cm^2)

Atmospheric radiation: Infrared radiation emitted by or being propagated through the atmosphere. It consists of both up flowing and down flowing components.

Atomizer: A device that produces a mechanical subdivision of a bulk liquid, as by spraying, sprinkling, or misting.

Atomizing carburetor: A type of carburetor for internal combustion engines, in which the liquid fuel is converted into a fine spray and mixed with the proper proportion of atmospheric air. Known also as Spray carburetor.

Atomizing nozzle: The spraying nozzle attached to the atomizing or spray carburetor of a gasoline motor.

Attachment: Any device used in connection with a machine in order to give a wider range or better quality of work.

Auger-feed mechanism: A feed and or distributing mechanism consisting of an auger, which causes a substance to flow evenly.

Auto ignition temperature: The minimum temperature required to initiate self-sustained combustion in a combustible fuel mixture in the absence of a source of ignition. Also known as self-ignition temperature.

Automatic cut off valve: An automatic device, which closes when the pressure of the fluid or gas reaches predetermined value.

Automatic declutching control: A control designed so that it will automatically interrupt power to a drive when the operators actuating force is removed.

Automatic tuber planter: A tuber planter with automatic feed mechanism with or without hand or automatic compensator.

Automobile: A self-propelled vehicle made to run on ordinary roads, transporting passengers & commodities.

Automobile chassis: An automobile frame, which with the wheels, power train, brakes, engine, and steering system are mounted.

Automotive air pollution: Evaporated, fully burnt and partially burnt fuel and other undesirable byproducts of combustion that escaped from a vehicle into the atmosphere, mainly carbon monoxide (CO), carbon dioxide (CO_2), hydrocarbons, nitrogen oxides (NO_x), sulfur oxides (SO_x) and particulates.

Available heat in drying air: The quantity of heat in air that may be utilized in evaporating water from the grain.

Average velocity of airflow through a product: The rate of air travel through product void space. It is determined by dividing the apparent velocity by the product void space.

Avogadro's law: Equal volume of all gasses, at the same temperature and pressure, contains equal number of molecules.

Avogadro's Number: The number of molecules in one mole of gas ($6.022.169 \times 10^{23}$ per mole). Under normal conditions (pressure of one standard atmosphere and temperature of 0° or 32°F), the volume occupied by one mole of gas is the same for all permanent gases (22,421 cubic centimeters or 22.42 liters). Named for Amedo Avogadro (1776 -1856), an Italian chemist who identified this relationship.

Axial: Pertaining to rotating on or about an axis.

Axial fan: A fan whose housing confines the gas flow to the direction along the rotating shaft at both the inlet and outlet.

Axial flow fan (axial flow blower): An appliance for producing airflow parallel to the blower shaft.

Axial flow pump: A pump having an axial flow or propeller type impeller; used when maximum head are desired. Also known as Propeller pump.

Axial force: A force, which acts uniformly over the section of a prismatic body so that its resultant coincides with the axis of the body.

Axis: a central line considered in relation of certain geometrical or mechanical relations. A point or line about which or on which a body revolves.

Axis of symmetry: An imaginary central line around which a symmetrically developed body is formed, and in which the center of gravity is found.

Axle: A shaft or device which carries the driving, traveling, or truck wheels of a vehicle, such as a tractor, trolley, automobile, locomotive, wagon, etc.

Axle load: The technically permissible axle load for each axle stated by the manufacture.

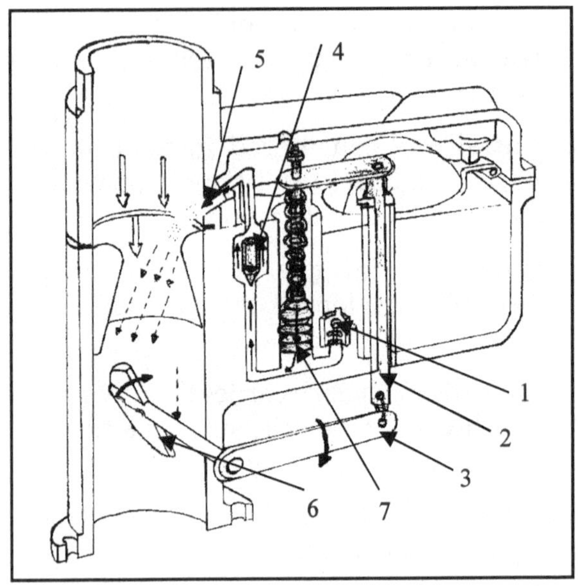

1. Fuel inlet valve 2. Throttle linkage connection 3. Throttle valve lever 4. Fuel outlet check valve 5. Accelerating pump jet 6. Butterfly/throttle valve 7.Plunger

Fig.1A Accelerating pump system of carburetor

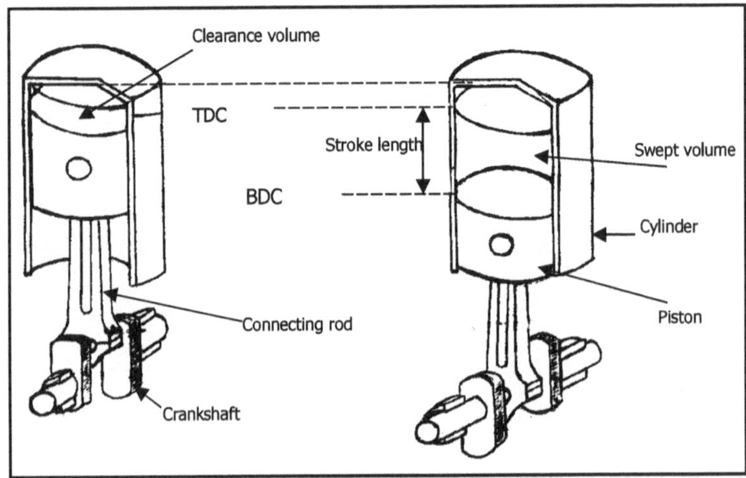

Fig. 2A Top Dead Center and Bottom Dead Center positions of piston

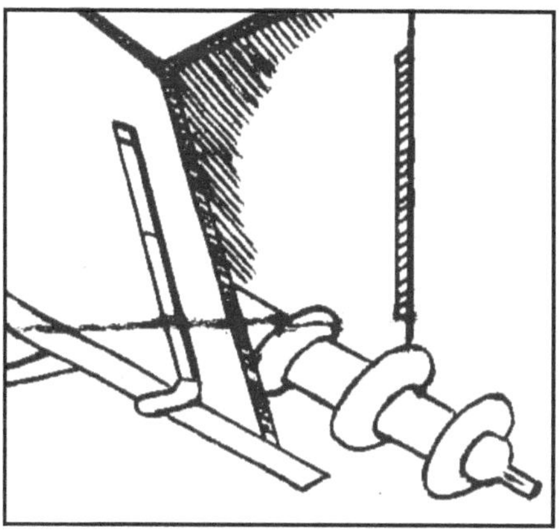

Fig. 3A Agitator feed mechanism

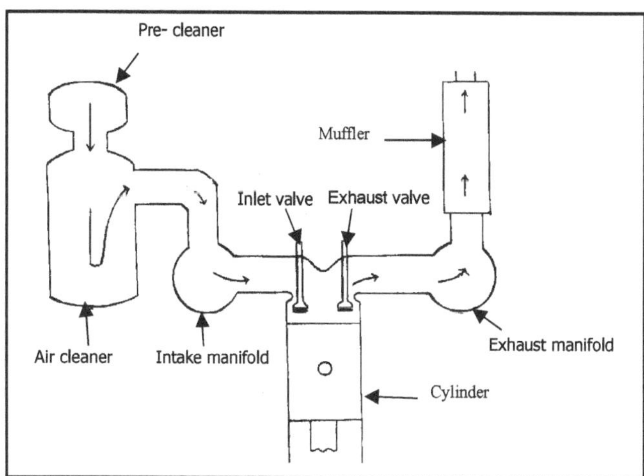

Fig. 4A Inlet and Exhaust system of CI engine

Fig. 5A Alternate plough

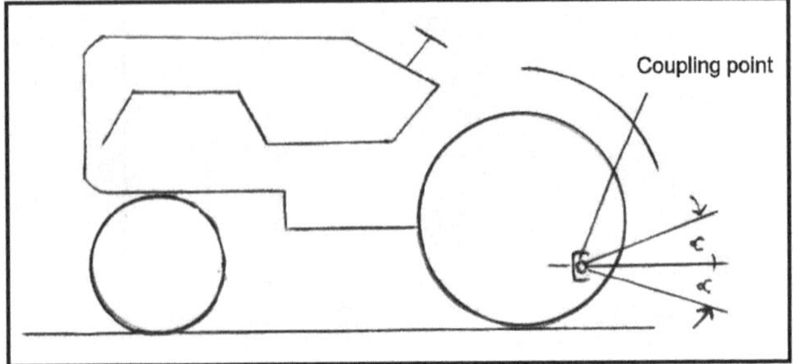

Fig. 6A Angle of pitch

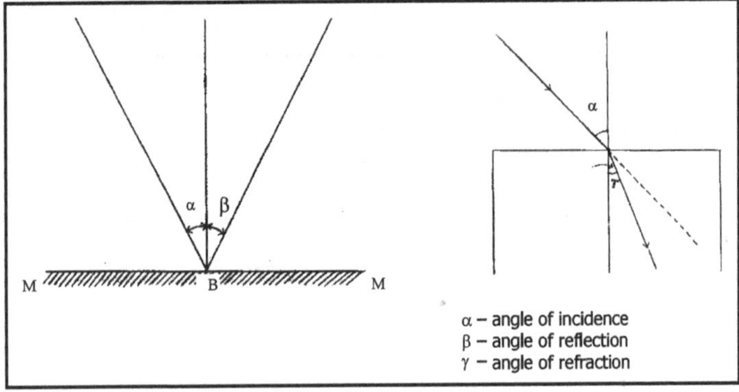

Fig. 7A Reflection and refraction of light

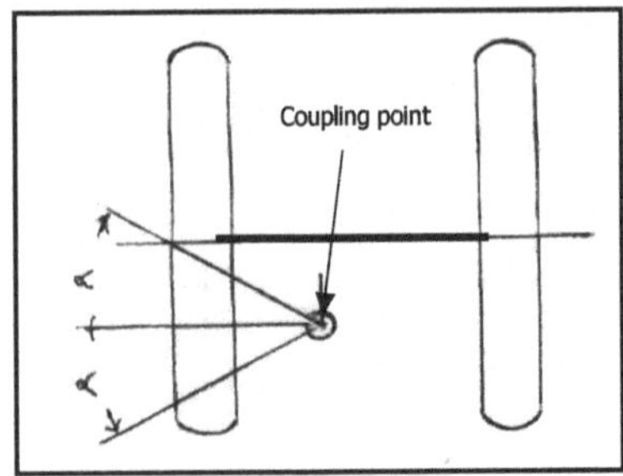

Fig. 8 A Angle of roll

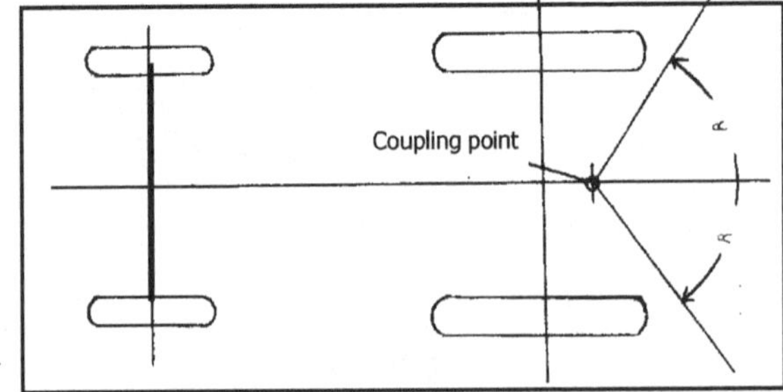

Fig. 9A Angle of yaw

Babbitt: 1. A tin base alloy containing 88% tin, 8% antimony and 4% copper. 2. In machinery, to line, bush, fill or face with Babbitt metal, e.g. to babbit a bearing.

Back furrow: A raised ridge left at the centre of the strip of land when ploughing is started from centre to side. *See* Fig. 1B.

Backfire: The accidental explosion of an overly rich mixture in the exhaust manifold of a spark-ignition engine. Backfire conditions can also develop if the premature ignition occurs near the fuel intake valve and the resultant flame travels back into the induction system.

Backlash: The clearance or play between two parts, such as meshed gears.

Backpressure: An opposing atmospheric force built up in the cylinders of an engine due to faulty timing, carbon deposit, dirty or fouled muffler, which prevents an engine from delivering its full power.

Baffle plate: 1. An obstruction for chocking or deflecting the flow of gases or sound. 2. In combine harvesters, it is an element placed near the rear beater to prevent grain from being thrown to the straw walkers.

Bakhar: An implement consisting of one or more blades attached to a beam or a frame used for shallow working of soil with minimum of soil inversion. It is also known as 'blade harrow'. *See* Fig. 2B.

Balance: It is a sensitive instrument for weighing.

Balance plough: A plough with bodies mounted facing one another on beams or frames, which are situated on opposite sides of transverse axis and are joined together at an obtuse angle. The plough is not turned at the end of the field, but is partially rotated about the axis and works to and fro with right and left bodies alternately in work.

Balance mass: The mass placed on the opposite side of the shaft of a certain rotating mass so as to balance the effect its centrifugal force. This process of providing the second mass in order to counteract the effect of the centrifugal force of the first mass is called as 'balancing of rotating masses'.

Balance wheel: A flywheel having the purpose to ensure smooth and regular motions in an engine or machine.

Balanced trailer: An agricultural trailer, of which total load is supported by its wheels when detached from towing tractor. This is the double or multiple axle trailer.

Balancing: In mechanics, the removal of irregularities in weight from the particular portions of a revolving body where are in excess, in order to equalize the strains upon it due to momentum and centrifugal force.

Balancing machine: A machine, which measures the 'out of balance' of a revolving piece and indicates the amount and location of the necessary correction.

Baler: A machine, used to collect the crop or straw from swaths or windrows, to compress it to rectangular shaped mass and to tie firmly with wire or twine for easy handling and storage.

Ball: It is a spherical metallic or non-metallic element used in valves, bearings etc.

Ball and socket joint: A joint in which a ball or spherical object is placed within a free motion in any direction within certain limits. A ball and socket mounting is usually applied to shafting supports to make them self-adjusting.

Ball bearing: An antifriction bearing permitting free motion between moving and fixed parts by means of balls confined between outer and inner rings.

Ball peen: It is the smaller or narrower end of a hammerhead, which is ball shaped (rounded).

Ball peen hammer: It is a commonly used hammer having one end of the head is ball shaped (rounded) for riveting and the other end is flat for striking a chisel, or other such works. *See* Fig. 3B.

Ball valve: A valve, in which the fluid flow is regulated by a ball moving in a spherical socket as a result of fluid pressure and the weight of the ball.

Ballast: Any mass that can be added or removed from a tractor for the purpose of changing traction and stability.

Band saw: A power operated wood working saw consisting basically of flexible band of steel having teeth on edge, running over two vertical pulleys, and operated under tension. *See* Fig. 4B.

Bar point share: It is a type of share in which point of the share is provided by an adjustable and replaceable bar. *See* Fig. 5B.

Bar: It is a metric unit of pressure, which is equal to one atmosphere. 1 Bar = 10^5 Pa = 1.019 kg/cm^2.

Barometer: An instrument for measuring atmospheric pressure.

Basil: The beveled edge of a drill or chisel.

Basin lister: It is an implement, used for forming basins continuously in the rows.

Batch: Quantity of the items/products taken on a repetitive basis, for certain treatment.

Batch manufacturing: The manufacture of parts in discrete runs or lots, generally interspersed with other production procedures.

Batch type furnace: A furnace used for heat treatment of materials, with or without direct firing. The loading and unloading operations are carried out through a single door or slot.

Battery: An energy storage device that produces electricity by means of chemical action. It consists of one or more electric cells connected as a unit.

Battery ignition system: The ignition system in which the initial current is supplied by the battery. This system does not generate its own voltage. It modifies the voltage taken from the battery and charging system and delivers it at the correct time to fire the spark plugs. The components of battery ignition system are spark plug, distributor, ignition coil, condenser, ignition switch, dynamo, battery etc. Refer spark plug, distributor, ignition coil, condenser (electronic), dynamo and battery. *See* Fig. 6B.

Bead lock rim or ring: A lock ring which forces the toe of the tyre bead against the rim, and holds it there against possible movement even when the tyre is deflated.

Beading: The process of forming of a bead on piece of work for stiffening.

Beading machine: A machine, which by means of rolls of different shapes and size impresses a bead in the metal in shape according to rolls used.

Beading tool: A tool used for finishing the riveted ends.

Beam (implement): A rigid member connecting an implement/machine with power source. *See* Fig. 2B.

Bearing metal: Antifriction and white metals such as brass, gunmetal, and various alloys, which are used for making or lining the bearings of journals.

Bearing pressure: A localized compressive stress at the surface of contact between two members of a machine.

Bearing spring: The spring, which carries the weight of the vehicle, and rests on the axel box preventing jerks and shocks.

Bearing surface: 1. The surface over which a load is distributed. 2.In machinery, the surfaces of bearing parts which are in mutual contacts.

Beater bar drum: A threshing cylinder having helically arranged rubber faced angle iron beater bars.

Bed planting: Planting on beds, which are separated by furrows.

Bed plate: In machinery, the foundation plate to which the frame of a machine is bolted. It is usually of a cast iron with suitable projections to which the different parts of the machine or engine are bolted. It is also called as base plate and sole plate.

Bedded land: The land, which is broken in alternate back furrows and dead furrows.

Bedding: A tillage operation, which places soil into a specific configuration.

Bell crank: A lever or crank having two arms, which meet at an angle of 90 degrees. *See* Fig. 7B.

Bellow: A mechanism that expands and contracts having rising and falling top sucking air through a valve and blowing it out through a tube.

Belt: A band of leather or other flexible substance, passing around two wheels, transmitting power from one to the other.

Belt clamp: A device for holding the two ends of a belt together while they are being connected.

Belt driven: Operated by a driving belt and pulley.

Belt pulley: A pulley driven by power unit to transmit power to another machine or machines by means of belt. *See* Fig. 8B.

Belt shifter: A flat strip of a metal or wood having shifter fingers or arms used for shifting a belt from tight to loose pulley and *vice-versa*.

Belt tightner: A device employed to maintain a uniform tension upon driving belts

Bench dog (wood working) **:** A peg of wood or metal inserted in a slot or hole near the end of a bench used to prevent a piece of work from slipping.

Bench lathe: Small lathe mounted on a workbench for handling the light works.

Benchmark: A permanent reference mark placed upon a stone pillar or similar object in order to serve as a reference point in survey works.

Bend: It is the characteristic of an object, such as a machine part, that is curved.

Bending machine: A machine for bending a metal or wooden part by pressure also known as bender.

Bending moment: The product of applied force and the perpendicular distance between the support and the point at which the force is applied.

Bendix drive: The pinion gear mounted on the starting motor shaft, which automatically engages the flywheel ring gear when speeded up and disengages when driven by the flywheel ring gear.

Bernoulli's theorem: The fundamental theorem of hydraulics, which states that the total energy (consisting of potential energy, kinetic energy and pressure energy) of flowing fluid remains constant.

Bevel: Any surface not at right angles to the rest of the piece.

Bevel gear: A pair of gears, used to connect two shafts for power transmission having their axes intersecting.

Bi fuel: A system, that can operate on two fuels, one at a time and not simultaneously, such as gasoline and CNG.

Bill hook: A hand tool used for lopping of branches and cleaning in plantation and estate. It may have single edge or double edge. *See* Fig. 9B.

Binary vapour cycle: Is a vapour cycle in which the condenser of the high-temperature cycle (also called the topping cycle) serves as the boiler of the low-temperature cycle (also called the bottoming cycle). That is, the heat output of the high-temperature cycle is used as the heat input to the low-temperature.

Black annealing: A method used to anneal sheets or other products when it is not essential that the product be free from scale and discoloration.

Black smith: A smith who forges metal by hand.

Blade: The flat, active working part of a tool, instrument, or device, as the blade of a knife, the blade of a propeller, etc.

Blade angle: The acute angle between the chord of a section of a propeller, or of a rotary wing system and a plane perpendicular to the axis of rotation.

Blade tip circle: The path described by the outermost point of the cutting means as it rotates about its shaft axis.

Blank: A piece cut from a flat sheet before any forming operation is taken place.

Bleeder valve: A small screw in the hydraulic brake wheel and master cylinders, which when removed, permits the air to bleed through the flexible tube from the cylinder to a vessel containing fluid.

Bleeding hydraulic brakes: Removing air from the fluid lines and cylinders of the hydraulic brake system.

Blending: The process of mixing two or more different products together such as grains and supplements to obtain desired food ratios, or of mixing different quantities of the same product with different moisture contents to obtain a final mass with a uniform moisture content.

Blind hole: A hole, which does not pass completely through a work piece.

Block (engine): The major casting of the engine holding the cylinders, crankshaft, and camshaft. When you look at an engine, most of what you see is the block.

Block brake: A brake, which consists of a block or shoe of wood bearing upon an iron or steel wheel.

Blower: A rotary device, which generates a draught of air.

Body (plough)**:** The main frame to which shoe, beam and handle are attached.

Body (nozzle)**:** The main component of the nozzle on which the other components are fitted.

Boiler: The closed container in which water is heated and steam is generated which is used for heating or as a power source for industrial plants, ships, household heating, etc.

Boiling point of water: The temperature at which water begins to boil, under normal conditions. It is equal to 100^0 C or 212^0 F.

Bonded strain gage: The strain gage in which the resistance element is a fine wire, usually in zigzag form, embedded in an insulating backing material, such as impregnated paper or plastic, which is cemented to the pressure sensing element.

Boom type sprinkler system: An elevated, cantilever sprinklers mounted on a central stand. The sprinkler boom rotates about a central pivot.

Boot (seed drill)**:** The part of the sowing machine, which conveys, seeds or fertilizer from the delivery tube to the furrow.

Bore: A hole or diameter of a hole or diameter of a piston.

Boring: Making or finishing circular holes in wood or metal.

Boroscope: Apparatus used for the detailed visual inspection of internal cylinder and tube surfaces, and allows for close inspection of affected areas.

Boss: The center or hub of a wheel.

Bottom dead centre (BDC)**:** The lowest position of the piston in the cylinder, at which it forms the largest volume in the cylinder. *Refer* Fig. 2A.

Bottoming cycle: A power cycle operating at low average temperatures that receives heat from a power cycle operating at higher average temperatures.

Boundary work (Pdv work)**:** The work associated with the expansion or compression of a gas in a piston-cylinder device. Boundary work is the area under the process curve on a P-V diagram, which is equal in magnitude to the work done during a quasi-equilibrium expansion or compression process of a closed system.

Boundary: 1. A limiting or dividing line or mark. 2. The real or imaginary surface that separates the system from its surroundings. The boundary of a system can be fixed or movable.

Bourdon tube: Named after the French inventor Eugene Bourdon, is a type of commonly used mechanical pressure measurement device which consists of a hollow metal tube bent like a hook whose end is closed and connected to a dial indicator needle.

Boyle's law (First gas law): Temperature remaining constant, the volume occupied by a given mass of gas varies inversely with its absolute pressure . Named for British physicist *Robert Boyle* (1627-1691).

Brake cylinder: The hydraulic cylinder within the brake drum used to carry the brake shoe setting pistons. The cylinder is usually open at each end and has two pistons moving in opposite directions as hydraulic brake fluid is forced between them.

Brake dynamometer: Used for measurement of brake horse power of the engine. Types of brake dynamometers are as 1. Prony barke, 2.Electric, and 3.Hydraulic.

Brake fluid: A mixture of glycerin, oils and additions of such composition, which will retain proper fluidity in the braking system under varying weather conditions and will withstand successive heating.

Brake hose: The flexible hose connection between the front wheel cylinder and the metallic tubing used to conduct the brake fluid to the front wheel cylinders.

Brake horse power (B.H.P.): The power developed by an engine in actual operation. B.H.P.= I.H.P. (Indicated horse power) - Frictional losses.

Brake lever: A lever to which brake rods are attached in a mechanical braking system. *See* Fig. 10B.

Brake lining: Woven material, mostly of asbestos or other heat-resistant material, applied to brake shoes and brake bands to produce friction. *See* Fig. 10B.

Brake pedal: The foot pedal, which controls the application of the brakes. *See* Fig. 10B.

Brake ratchet: A notched quadrant, which engages a pawl attached to the brake lever permitting a setting of the brake lever in a fixed position.

Brake rod: A rod terminating with an eye or clevis, forming a part of the linkage in a mechanical braking system. *See* Fig. 10B.

Brake shoe: A metal casting fitted to the curve of a wheel and pressed against the wheel to secure braking action. *See* Fig. 10B.

Brake shoe clearance: The space between the lining on the brake shoe and the inner surface of the brake drum.

Brake testing machine: A machine on which the results of brake adjustment can be tested.

Brake thermal efficiency: The ratio of brake power output to power input.

Brake drum: The cast iron or pressed steel pan like part used for receiving the brake-shoe application when the brakes are set. It is mounted on the wheel hub and turns with the wheel.

Braking surface: The surface of contact between the stationary and moving parts of a brake.

Braking system: A combination of one or more brakes and the related means of operation and control.

Bran: The powdered material obtained from the outermost part of the brown rice.

Brazing: The process of joining two or more pieces of metal with an alloy. The joints are cleaned, bound with wire and sprinkled with borax, then heated until the alloy melts.

Brazing metal: An alloy formed from the two parts of tin and ninety-eight parts of copper.

Breaker mouldboard or Sod mouldboard: A long mouldboard with a gentle curvature which lifts and inverts unbroken furrow slice. It is suitable for the land having full of grasses. *See* Fig. 11B.

Breaker (spark distributor)**:** The device, consisting of the contact points, the cam and the movable arm within the spark distributor used for opening (breaking) the 6/12 volts circuit.

Breaking strength: The ultimate resistance to rupture of a piece of material of specified size.

British thermal unit (BTU): The energy unit in the English system needed to raise the temperature of 1 lb of water at 68 °F by 1°F.

Brittleness: The property of breaking a material with little permanent distortion.

Broadcast tillage: Coverage of an entire area as contrasted to a partial coverage as in bands or strips. It is also known as overall tillage.

Broadcasting: The process of scattering agriculture inputs, such as seed, fertilizer and manure on the surface of the soil.

Broken rice: Rice kernels, which are broken and whose length is less than three forth of the original length.

Brown rice: Paddy, from which only husk has been removed. It is also known as dehusked rice.

Brush feed mechanism: A feed mechanism, in which a rotating brush regulates the flow of seed from the hopper.

Brush hog (bush hog)**:** A rotary flailing blade mower.

Brush (electric)**:** An electrical conductor, usually carbon that picks up or delivers electrical current from a stationary or a moving part.

Brush holder (electric)**:** The socket in which the carbon brush rests and moves.

Bucking: The process of cross cutting the felled or uprooted trees into lengths.

Budding knife: A knife to remove the bud from the scion, to cut the stock, slit and raises the bark to insert the bud.

Budding tape: Tape used in holding the buds in their position after budding operation.

Buff: An appliance for polishing and finishing metallic surfaces, which consists of a large number of muslin disks fastened together to form a polishing wheel.

Buffing: A term applied for roughing up with an emery wheel or polishing down with a soft fabric wheel. It is also applied on leather to polish the surface or take off the grains on fancy leathers.

Bulldozer: A blade located on front of a tractor (usually a crawler) to push dirt.

Bulldozing: The pushing and rolling of land by inclined blade.

Bunching: Gathering and arranging the tree or parts of tree in bunches to facilitate loading and transportation.

Bund former: An implement, which gathers soil and forms a bund. *See* Fig. 12B.

Burning speed (flame)**:** The speed at which a flame travels through a combustible gas mixture. It is different from the flame speed.

Burr: The ragged or turned down edges of a piece of metal resulting from grinding, cutting or punching.

Bush: To fit a hollow sleeve into the covering part which may serve as a bearing. When wear occurs, a new bushing may be applied, avoiding the necessity of replacing an entire part.

Butane: A type of petroleum gas which is in liquid form below 32 °F (0°C) at atmospheric pressure.

Butt joint: A Joint, in which the main plates are kept in alignment butting (i.e. touching) each other and a cover plate (i.e. strap) is placed either on one side or on both sides of the main plates. The cover plate is then riveted together with main plates.

Butterfly valve: A winged damper in the carburetor supported on a fulcrum, movement of which controls the passage of gas or air through an opening. *Refer* Fig. 1A.

Butt-welding: A method of welding by electric arcs in which, two pieces to be connected are welded directly at their ends without overlapping each other.

By pass: An alternate path, which allows all or a part of the fluid (gaseous or liquid), delivered by the pump to return to the tank.

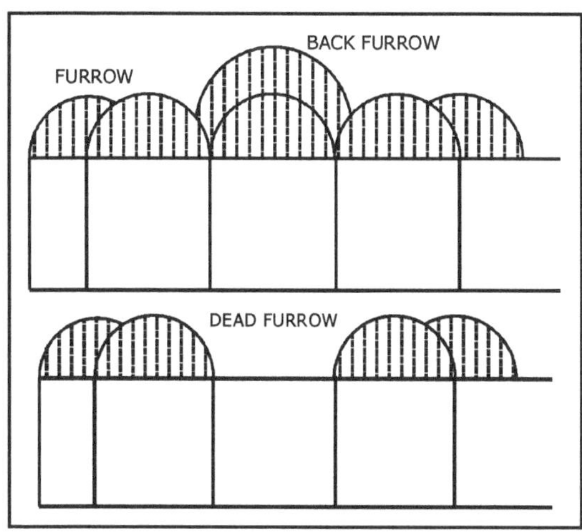

Fig. 1B Plough furrow

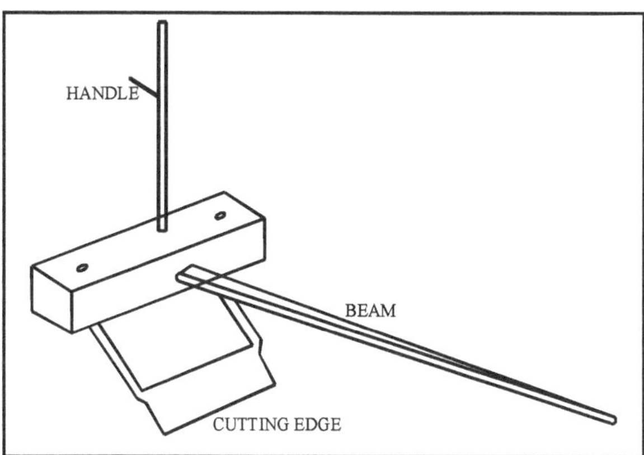

Fig. 2B Plough furrow

Fig. 3B Ball peen hammer

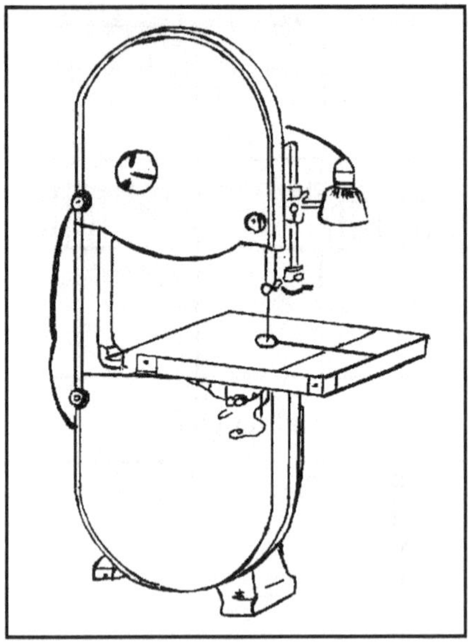

Fig. 4B Band saw

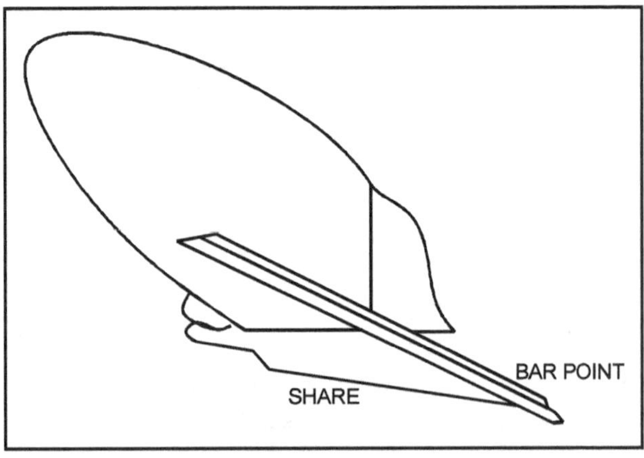

Fig. 5B Bar point share

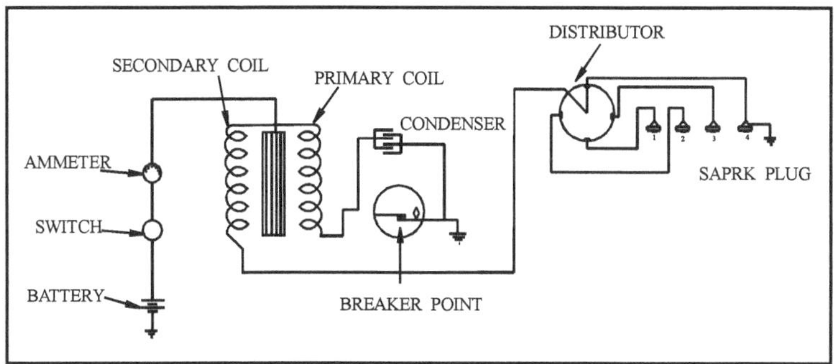

Fig. 6B Battery ignition system

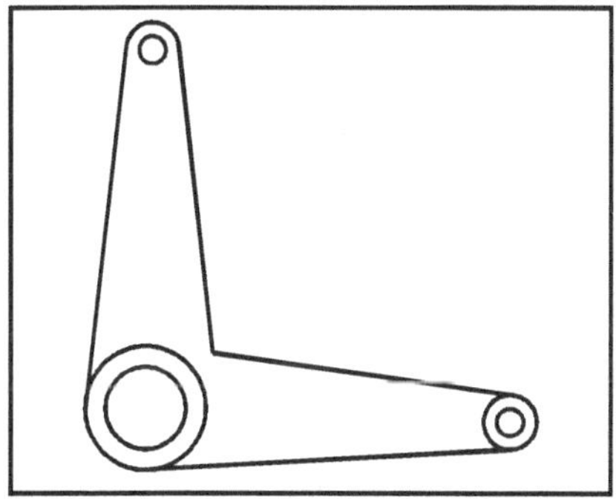

Fig. 7B Bell crank lever

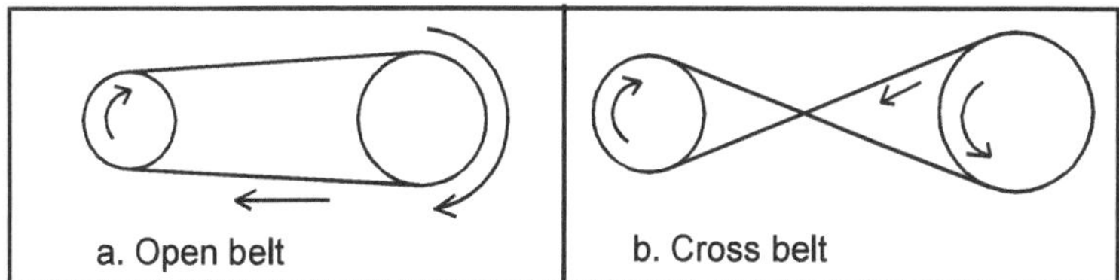

Fig. 8B Belt pulley

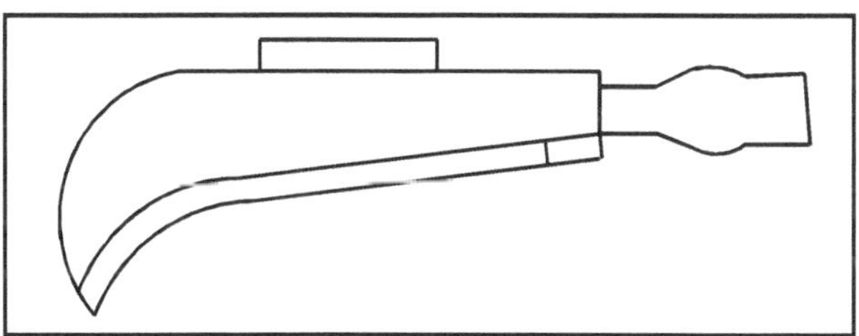

Fig. 9B Bill hook

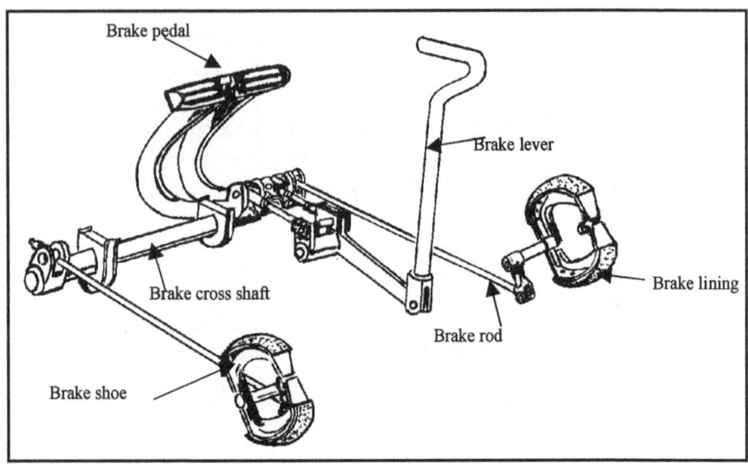

Fig. 10B Mechanical brake system

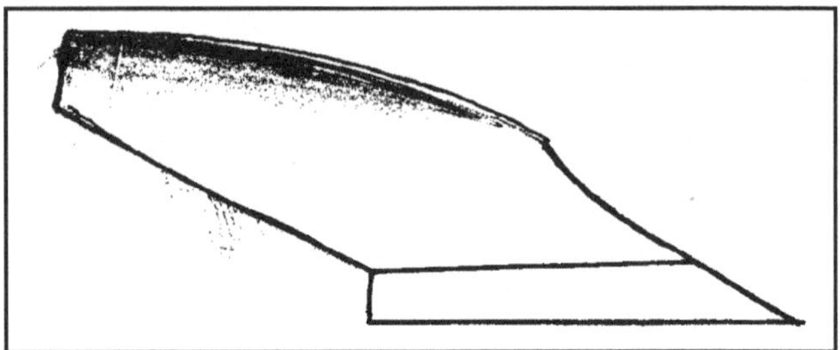

Fig. 11B Breaker (sod) mouldboard

Fig. 12B Bund former

Cable conveyor: A power conveyor in which a trolley runs on a flexible, torque transmitting cable that has helical threads.

Cable: 1. A general term applied to rope or chain used for engineering purposes. 2.The standard wire with insulating cover used for conducting electrical current.

Cage harrow: A rolling harrow with horizontal shafts each fitted with star shaped tines joined by rods forming the lines of cylinder.

Cage roller: Roller consisting of a cage of metal bars arranged so as to form the shape of a cylinder. *See* Fig. 1C.

Cage wheel: A wheel or an attachment to a wheel, used in wet land seed bed preparation; having spaced cross bars for reducing ground pressure and improving traction of a tractor, or power tiller. *See* Fig. 2C.

Cage wheel, full: A cage wheel used in place of pneumatic wheels (tyres).

Cage wheel, half: A cage wheel being used in conjunction with pneumatic wheel.

Calibrate: To determine or adjust the graduation of scale of any measuring instrument.

Calking: Making a joint or seam watertight or steam-tight by filling it in with rust cement or by closing the joint by means of a calking tool.

Calking tool: A piece of steel with a blunt end formed somewhat like a chisel and used for closing up the joints of boilerplates. It is driven against them by sharp blows of a hammer.

Calorie: The metric measurement of the amount of heat required to raise the temperature of one gram of water from zero degree to one degree centigrade. 1 Calorie = 4.184 joules.

Calorific value: Heat liberated by combustion of fuel per its unit mass. It is expressed in kcal per kg.

Cam: A turning or sliding piece which, by the shape of its periphery or a groove in its surface, imparts variable or intermittent motion to, or receives such motion from a rod, lever, or block brought into sliding or rolling contact with it. *See* Fig. 3C.

Cam drive: A type of motion obtained by cams through which a certain motion is made to take place in exact time or relation to some other motion, as in the camshaft of an automobile.

Cam dwell: The part of cam surface between the opening and closing acceleration sections.

Cam follower: The output link of a cam mechanism. *See* Fig. 3C.

Cam governor: A controlling device used in Otto cycle gas engines having a stepped or differential cam giving three or four grades of valve lifts.

Cam lifter: Adjustable metal parts fitted between the camshaft and the valve stems.

Cam mechanism: A mechanical linkage whose purpose is to produce, by means of a contoured cam surface, a prescribed motion of the output link.

Cam profile: The shape of the contoured cam surface by mean of which motion is communicated to the follower. *See* Fig. 3C.

Cam wheel: A wheel with a projection or projections either on the periphery or face, adapted to give motion to another object against which it impinges, by sliding contact.

Camber: The feature of front axle design, which brings front wheel closer together at the bottom than at the top, for ease in steering. It is the difference between measurements of space between top and bottom of front wheel.

Camber angle: 1. The angle between centre line of the tyre and the vertical line. Or. 2. The inclination from the vertical of the steerable wheels of an automobile. *See* Fig. 5C.

Cambridge roller: Roller comprising a number of thick discs each having the form of two frusta of a cone with a straight periphery, which are jointed, base to base placed on same shaft without gap. *See* Fig. 6C.

Camfer: A level or taper at the edge of a hole.

Camshaft: A shaft carrying a series of cams which form integral parts of the shaft and are so spaced as to control the relative operation of the automobile engine valves. *See* Fig. 4C.

Cantilever: A beam fixed at one end and loaded at the other, or loaded uniformly.

Capacitor: An energy storage device that stores electrical energy in the form of an electrical charge. A capacitor consists of two metal plates with an insulating dielectric between them. Also refer condenser (electronics).

Capacity: The ability to perform work or hold (electric charge).

Carbon (C): A non-metallic element placed at No. 6 in the periodic table, constituent of all organic compounds and also occur in the combined form in many inorganic substances as carbon dioxide, limestone etc. It is a primary constituent of all hydrocarbon fuels. Carbon is routinely left as a black deposit on engine parts such as piston, rings, valves etc. by the combustion of fuel.

Carbon arc welding: An arc welding process wherein a carbon or graphite electrode is used with or without the use of filler metal.

Carbon deposit: An accumulation of carbon in the combustion chamber and some times around valves and piston rings. Poor quality of gasoline and lubricating oil hastens such actions.

Carbon dioxide (CO_2): A colourless, odorless gas that results from the complete combustion of carbon with oxygen. Carbon dioxide is a greenhouse gas and is a major contributor to the greenhouse effect.

Carbon knock: Premature ignition resulting in knocking or pinging in an internal combustion engine caused when the accumulation of carbon produces overheating in the cylinder.

Carbon monoxide (CO): A pollutant from engine exhaust that is a colorless, odorless, tasteless, poisonous gas that results from incomplete combustion of carbon with oxygen.

Carbon piles: Carbon plates forced together to act as a resistance unit in a high discharge electrical test set.

Carbonization: The oldest method of producing a hard surface on steel. In this process steel is heated to the temperature of about 1094^{0}C by packing in a container with carbonizing material such as charcoal, wood , bone etc with the compounds such as carbonates of barium, calcium etc. (called as energizer). Steel at temperature close to and above its critical temperature, has an affinity of carbon. The carbon thus enters the metal to form a solid solution with iron and converts the outer surface into a high carbon steel.

Carboy: A large glass bottle or container used for handling electrolyte.

Carburetion: The actions that take place in the carburetor; converting liquid fuel to vapor and mixing it with air to form a combustible mixture.

Carburetor: A device in an engine fuel system that converts liquid fuel into vapour and mixing it with air in such proportions as to form the most efficient combustible mixture and supplies it to the intake manifold for varied speed and load conditions of an engine. *See* Fig. 7C.

Carburetor barrel: The metal portion of the carburetor surrounding the suction air chamber forming a well or barrel for the full length from choke to flange.

Carburetor bowl or float chamber: The reservoir part of the carburetor, which holds liquid gasoline or fuel. *See* Fig. 7C.

Carburetor chock: A butterfly valve fitted in the air passage of the carburetor for restricting the supply of air so that mixture becomes rich which helps in easy starting of an engine. *See* Fig. 7C.

Carburetor compensating jet: It maintains correct proportion of air and fuel

for different load and speed conditions.

Carburetor economizer: A device installed in a carburetor to control the amount of fuel used under certain conditions.

Carburetor float: An airtight metal container, which floats on the surface of the fuel in the bowl of the carburetor and controls the flow of gasoline coming from the main fuel line. *See* Fig. 7C.

Carburetor idling jet: It is a passage on the manifold side of the butterfly valve. When butterfly valve is closed, engine suction pressure takes the only required fuel for running the engine at idling speed through this jet. Also see idling jet.

Carburetor main jet : A fixed or adjustable type small opening through which fuel passes from the float chamber to throat of the carburetor in the form of spray. Small stationary (single cylinder or multicylinder) engines are equipped with fixed type and tractor engines are equipped with adjustable type of jet.

Carburetor throttle: A butterfly valve between the mixing chamber of the carburetor and inlet inlet manifold of engine to regulate the quantity of change operated by hand lever, foot lever or a governor. See also throttle and throttle valve. *See* Fig. 7C.

Carburetor icing: The formation of ice in an engine carburetor as a consequence of expansive cooling and evaporation of gasoline.

Carnot cycle: A hypothetical cycle consisting of four reversible processes – two isothermal and two adiabatic, and can be executed either in a closed or a steady flow system. French engineer Sadi Carnot first proposed it in 1824.

Carnot efficiency: Is the highest efficiency of a heat engine operating between the two thermal energy reservoirs at temperatures T_L (lower) and T_H (higher) equal to $1 - T_L / T_H$.

Carnot engine: An ideal, frictionless engine, which operates on the Carnot cycle.

Carnot's principle: The efficiency of an irreversible heat engine is always less than that of a reversible, which are operating between the same two reservoirs.

Carriage (lathe m/c): That part of a lathe, which rides on the ways between the headstock and tailstock, and through which the feed may be controlled, both longitudinal and transverse.

Casing: The outer enclosure surrounding any machine part or mechanism.

Caster angle: The angle at which the kingpin of front axle is set with reference to a vertical line. *See* Fig. 8C.

Caster: The effect secured by setting top of the front axle kingpin slightly back of the bottom. This causes the front wheels to align themselves with the direction in which the vehicle is moving.

Casting: Metal part made by pouring the metal, when liquid, into a mold.

Cast iron: Is pig iron remelted and thereby refined together with definite amount of limestone, steel scrap and spoiled casting in a cupola or other form of remelting furnace and poured in to suitable moulds of required shape. It contains about 2 to 4 per cent of carbon, a small per cent of silicon, sulphur, phosphorus and manganese and certain amount of alloying elements, e.g. nickel chromium, molybdenum, copper and vanadium.

Catalyst: A substance that can speed or slow a chemical reaction between substances, without itself being consumed by the reaction. Platinum is a typical catalyst.

Caterpillar: A vehicle, such as a tractor or army tank, which runs on two endless belts, one on each side, consisting of flat treads and kept in motion by toothed driving wheels.

Cathode: Electrically positive electrode at which gain of electrons (reduction) takes place

Cation: A positively charged ion.

Caulking: In general, making a joint tight or leak proof by forcing plastic material between parts that are not tightly fitted.

Cell feed mechanism : A feed mechanism in which seeds are collected and delivered by a series of equally spaced cell (apertures or indentation) on the periphery of circular plate or wheel. *See* Fig. 10C.

Celsius: Metric temperature scale and unit of temperature (°C). Named for Swedish astronomer Anders Celsius (1701:1744), who invented first thermometer in 1743 having 100^0 as the freezing point of water, and 0° as the boiling point, the reverse of the modern Celsius scale.

Centre: A fixed point about which the radius of a circle or of an arc moves.

Centre of gravity: The point about which a body will be balanced when placed in any position.

Centre punch: A steel punch with one end ground to a point; used for laying out work. *See* Fig. 9C.

Centimeter: A measure of length in metric system equal to the 100th part of a meter. 1 m = 100 cm=39.37 in.

Centre of resistance: The point at which the resultant of all the vertical and horizontal forces act.

Centrifugal: Proceeding from the centre.

Centrifugal fan (centrifugal blower) **:** An appliance for producing airflow at right angles to the blower shaft.

Centrifugal fertilizer distributor: A fertilizer distributor for solid fertilizers in which the fertilizer is broadcasted by centrifugal force.

Centrifugal governor: It works on the principle as – a weight made to rotate about a point, it tends to fly away in tangential direction from the centre of rotation. A throttle valve connected to the flying weights through the linkage can be made to operate.

Centrifugal pump: A non-volumetric pump in which the liquid energy is increased by one or more impellers.

Centrifugal spark advance: The timer-distributor governor by which the spark is advanced or retarded as the engine speed varies.

Centrifugal sprayer: A sprayer used for centrifugal spraying.

Centrifugal spraying: A spraying performed by the use of centrifugal force imparted to the liquid by mechanical rotational energy such as spinning disc.

Centrifugal type supercharger: A high-speed rotary blower equipped with one or more multiblade impellers, which, through centrifugal action compresses the air or mixture in the induction system.

Centrifugal-feed drill: A seed drill in which seeds are metered by centrifugal force.

Cetane number: The percentage of cetane in a mixture of cetane $C_{16}H_{34}$) and alphamethyl napthelene ($C_{11}H_{10}$) that produces the same knocking effect as the fuel under test. Diesel fuels are rated according to cetane number; higher the cetane number better is the ignition quality of fuel.

Chaff cutter: A machine with blades for cutting the fodder.

Chaffer: The upper sieve on which grain and chaff mixture falls from stepped grain bed for initial cleaning .The sieve is oscillated so as to toss the chaff and un-threshed material rearward while the grains sift through chaffer openings.

Chaffer extension: The bars provided at the rearward of chaffer to receive the material passed by chaffer for separating unthreshed grains chaff and remaining grain to the tailing auger.

Chaffing: The process of pneumatic cleaning of grain.

Chain brake: Device for stopping and locking the chain, which is activated manually and released automatically when kick back occurs.

Chain catcher: Device for restraining the chain if it breaks or degrooves.

Chain drive: The type of drive in which the power transmission is by chain belting.

Chain harrow: A harrow consisting of a frameless network of links.

Chain pitch: The arithmetic mean of the two distances between two adjacent rivets.

Chain pulley: A sheave wheel or pulley with depression to engage the links of a chain.

Chamfer: To bevel a sharp edge of a machined part.

Chamfer angle: The angle that a beveled surface makes with one of the original surfaces.

Chamfering: Machining operations to produce a beveled edge. Also known as beveling.

Charcoal iron: A very superior quality of iron, which is smelted with wood charcoal as fuel. Its quality lies in its freedom from sulphur.

Charging rate: The electric charge (number of amperes) flowing to the battery charging per unit time.

Charles' law (second gas law): Pressure remaining constant, the volume of a gas varies directly with temperature. Named for French scientist Jacques A.C. Charles (1746:1823), who reached this conclusion in 1787.

Check: To verify that a component, system or measurement complies with specifications.

Chassis: The unit, which includes all parts of the automobile except the body.

Check nut: Commonly called a lock nut. A nut screwed down upon another nut causing a binding on the threads.

Check valve: A valve that opens to permit the passage of air or fluid in one direction only, or operates to prevent (check) some undesirable action.

Check wire (check row planter)**:** A wire or rope with equally spaced buttons or knots tied between two stalks at the ends of the field so as to actuate the check row planter.

Check-row planting: The process of planting in which row to row and plant-to-plant distance are uniform and plants across the rows are also in line.

Chemical energy: It is the internal energy associated with the atomic bonds in a molecule.

Chemical equilibrium reaction: Is a chemical reactions in which the reactants are depleted at exactly the same rate as they are replenished from the products by the reverse reaction. At equilibrium the reaction proceeds in both directions at the same rate.

Chemical equilibrium: Is established in a system when its chemical composition does not change with time.

Chemical formula: The representation of the chemical composition of a molecular compound or substance according to its constituent atoms.

Chemical reaction: The formation of one or more new substances when two or more substances are brought together.

Chimney: A vertical flue for drawing off the products of combustion from a stove, a furnace, a fireplace, or some other smoke and gas-producing source.

Chip: 1. To remove particles as with a hammer and chisel. 2. A small (metallic or non metallic) piece broken or scaled off.

Chipper (tree) **:** Machine design to chip trees or parts of trees.

Chipping: Converting logs or billets in to small pieces for further processing.

Chisel: Tool of great variety whose cutting principle is that of the wedge.

Chisel plough: A plough used to cut through hard soils by means of one or a number of narrow tines. *See* Fig. 11C.

Chiseling: A tillage operation in which a narrow tool is used to break up hardpan in the soil. It is usually performed at a depth greater than the normal ploughing depth.

Chromium (Cr): A grayish-white metallic element having specific gravity 6.50 and melting point 2939^{0}F. It is used in alloy steel and in plating.

Chuck: A device for holding a rotating tools or work during an operation.

Chute: An inclined trough or tube to guide sliding objects from a higher to a lower level.

Chute fed machine : A machine, in which the feeding of the input material is done through a chute.

Circuit: An arranged path through which electric current flows.

Circular harrow: A rolling harrow consisting of circular frame fitted with teeth and turning about shaft slightly inclined toward the vertical.

Circular pitch: The distance from the center of one tooth to the center of the next, measured on the pitch line.

Circular saw: A saw whose teeth are spaced around the edge of a circular disk running upon a central arbor. *See* Fig. 12C.

Circumscribe: To enclose within certain lines or boundaries. To draw around or outside of.

Clamp: A frequently used form of clamp shaped like the letter 'C' pressure is obtained by means of a thumb screw. *See* Fig. 13C.

Clamp type transplanter: A transplanter, the planting mechanism of which consists of plates, which carry clamping devices to hold seedlings.

Clausius statement of the second law: It is impossible to construct a device that operates in a cycle and produces no effect other than the transfer of heat from a lower-temperature body to a higher-temperature body. It is first stated by the German physicist R. J. E. Clausius (1822-1888)

Claw coupling: A loose coupling used in cases where shafts require instant connections. It is somewhat like a flange coupling, but instead has projections or claws cast on each face, which engage in corresponding recesses in the opposite faces. *See* Fig. 14C.

Cleaning (grains) : The operation of isolating the desired grain from chaff, small debris, and incompletely threshed and completely unthreshed grains.

Cleaning sieve (grains) : One sieve or set of sieves, which separate chaff and other undesirable material elements from grain after being received from chaffer and chaffer extension.

Clearance height of unloader (combine harvester) : The vertical distance from plane on which the combine is standing to point on the underside of the unloader at horizontal distance of 1000 mm from the lowermost point of the discharge opening expressed in millimeters.

Clearance volume: is the minimum volume formed in the cylinder when the piston is at top dead centre.

Cleat: A flat or angle strip fitted to the rim of a steel wheel.

Cleated wheel: A steel wheel fitted with cleats.

Clockwise: The direction in which the hands of a clock move, with a right-hand motion.

Clods: Soil blocks or masses those are cut, sheared or broken loose by tillage tools.

Clod crusher: A rolling harrow with horizontal shafts fitted with star shaped tines

Clod strength: The mechanical strength of individual soil clod.

Clog: To obstruct, hinder, or chock up; e.g. the stoppage of flow through a pipe by an accumulation of foreign matter.

Closed system (control mass) **:** Consisted of a fixed amount of mass, and no mass can cross its boundary. But energy, in the form of heat or work, can cross the boundary.

Clutch: Device for engaging and disengaging a driven member to and from a rotating source of power. Clutches are of two types: 1. Friction clutch and 2. Dog clutch. Refer friction clutch and dog clutch. *See* Fig. 15C.

Clutch facing: The lining of friction material attached to the driven clutch plate.

Clutch fingers (release levers): Small levers in the clutch pressure-plate assembly operated by the release bearings which compress the clutch springs to remove the pressure from the driven plate, thus bringing the clutch into the off position. *See* Fig. 15C.

Clutch fork (Yoke)**:** The Y shaped clutch lever used for throwing out the clutch.

Clutch pedal: The left foot pedal, which connects and disconnects the clutch. *See* Fig. 15C.

Clutch pedal adjustment: Adjusting the linkage between the bottom of clutch pedal and the outer end of the throw-out yoke lever to provide the proper amount of clutch release when the clutch pedal is depressed.

Clutch pressure plate: Part of the clutch assembly consisting of a metal ring mounted on levers. Helical springs exert a pressure in turn passes to the clutch disk and to the machined face of the flywheel, providing the necessary friction to transmit power from engine to transmission. *See* Fig. 15C.

Clutch spring (pressure spring): The spring or springs used to force the pressure plate against the driven disc or clutch plate. *See* Fig. 15C.

Clutch vibration dampener: A flexible power-transmitting unit between the driven clutch plate and its hub (which fits onto the clutch shaft). Sometimes the dampener may be of rubber but more frequently it is in the form of coil springs.

Coal: A solid, brown to black, inflammable substance dug from the earth. It is formed from prehistoric vegetable deposites. Its main use is as fuel.

Coal gas: Gas produced by the destructive distillation of bituminous coal. It is an illuminating gas.

Coalescing filter: A filter designed to separate liquid from gas.

Coarse spray: The distribution of droplets with a volume mean diameter (VMD) value of more than 400 μm.

Coefficient of expansion: The factor, which expresses the change in length per unit length of any material for each degree of temperature.

Coefficient of friction: The ratio of the force required to move a body over a horizontal surface, to the weight of the body moved.

Coefficient of rolling friction: The ratio of the frictional force, parallel to the surface of contact, opposing the motion of a body rolling over another, to the force normal to the surface of contact, with which the bodies press against each other.

Coefficient of static friction: The ratio of the maximum possible frictional force, parallel to the surface of contact, which acts to maintain two bodies in contact and at rest to the force, normal to the surface of contact, with which the bodies press against each other. It is also known as coefficient of friction at rest.

Cold drawn: The process of production of metal into its final form by drawing through dies while cold.

Collars: A round ring on shaft or stud to prevent sliding.

Collet: A clamping ring or holding device. In the machine shops, this term is applied to sockets for tapered shank drills, and reducing sleeves and bushings of various types.

Combination of splash and forced feed lubrication system: Forced feed system is used for lubrication of engine components as main bearing, camshaft bearing and connecting rod bearing, which are subjected to very heavy load. While splash is used for lubrication of the rest of parts as cylinder liners, tappets, cams etc.

Combine harvester thresher or combine harvester: A machine designed for harvesting, threshing, separating, cleaning and collecting grain while moving through standing crop. Bagging arrangement may be provided with a pickup attachment, it may be used for handling the crop that has been swathed. It may be of self propelled type or tractor operated type.

Combine height: The vertical distance from the horizontal plane on which the combine is standing to the highest point of the combine expressed in millimeters.

Combine length: The overall length from the foremost point to the rearmost point of the combine equipped for field operation measured parallel to the longitudinal centerline of the combine express in millimeter.

Combine mass: The mass of the complete combine equipped for field operation expressed in kilogram.

Combine tuber planter: A machine to plant potatoes and apply fertilizer in one operation.

Combine width: The overall width measured horizontally covering outer extreme of combine expressed in millimeters.

Combined tillage operations: Operations simultaneously utilizing two or more different types of tillage tools or implements (subsoiler, lister, lister-planter or plough planter combinations) to simplify control or reduce the number of operations over the field.

Combustion: A chemical reaction during which a fuel is oxidized and heat energy is released.

Combustion air: Dry air, which can be approximated as 21 percent oxygen and 79 percent nitrogen by mole numbers. Therefore, each mole of oxygen entering a combustion chamber will be accompanied by 0.79/0.21 = 3.76 mol of nitrogen. To supply one mole of oxygen to a combustion process, 4.76 mol of combustion air are required.

Combustion chamber: The space between the top of the piston and the cylinder head when the piston is at top dead centre, in which the air fuel mixture is burned.

Combustion chamber volume: The volume of the combustion chamber when the piston is at top dead center.

Combustion efficiency: Ratio of the amount of heat released during combustion divided by the heating value of the fuel. A combustion efficiency of 100 percent indicates that the fuel is burned completely and the burnt gases leave the combustion chamber at room temperature, and thus the amount of heat released during a combustion process is equal to the heating value of the fuel.

Complete combustion: A combustion process in which all the combustible components of a fuel are burned completely. All the carbon in the fuel burns to CO_2, all the hydrogen burns to H_2O, and all the sulphur (if any) burns to SO_2.

Compressed hydrogen gas (CHG): Compressed hydrogen gas is hydrogen compressed to a high-pressure and stored at ambient temperature.

Compressed natural gas (CNG): Mixtures of hydrocarbon gases and vapors, consisting principally of methane in gaseous form that has been compressed for use as a vehicular fuel.

Compression (gas)**:** The process, in which the application of pressure to the gas reduces volume tending to increase density and temperature

Compression ignition (CI) : Ignition of fuel charge caused by the heat of compression. The compression ignition temperature for internal combustion engine varies from 500^0 to 900^0 F and compression pressure for easy ignition is 35 to 45 kg/cm^2.

Compression ratio (CR): The ratio of the volume of the cylinder when the piston is at bottom dead centre to the clearance volume. CR for diesel engine (CI) varies from 14:1 to 22:1 and that of for carburetor type engine (SI) varies from 4:1 to 8:1.

Compression ring: The upper ring or rings on a piston, designed to hold the compression in the combustion chamber and prevent blow by.

Compression sprayer: A hydraulic sprayer in which the liquid pressure is obtained by means of a previously compressed gas.

Compression springs: A helical spring, which is designed to operate under pressure, therefore tending to shorten when in action.

Compression stroke: The piston movement from bottom dead center to top dead center immediately following the intake stroke, during which both the intake and exhaust valves are closed while the air charge in the cylinder is compressed. *Refer* Fig. 9F.

Compressor: Equipment that pressurizes air, gas etc. into a compressed state.

Computer: A programmable electronic device that can store, retrieve and process data.

Concave (thresher)**:** A concave shaped metal grating partly surrounding the cylinder against which the cylinder rubs the grain from the ear heads and through which the grains fall on the sieve.

Concave grate area: The portion of the concave area, which is sievable for separation. This area shall be calculated using the outside dimension of the sievable surface and expressed in square millimeter.

Concave grate extension area: The product of concave grates extension width and concave length expressed in square millimeters.

Concave grate extension: A sievable element approximately concentric with the cylinder and generally forming an extension of the concave contour and helps in separating and bringing loose straw-to-straw walker.

Concave length: The minimum distance between the two panels of the combine in which the concave is mounted expressed in millimeter.

Concave width: The distance from the front of the first bar to the rear of the last bar, measured along the contour formed by the inner surface of the concave bars expressed in millimeters.

Concavity (disc)**:** The depth measured at the lowest point at the centre of the disc by placing concave side on a flat surface.

Condensation: The process of a vapour becoming a liquid; the reverse of evaporation due to temperature or pressure changes.

Condenser (electronics)**:** The device for storing electrical charges, consists of alternate sheets of conductor (often tin foil), separated by an insulator or dielectric (often paper). It is also called as 'capacitor'.

Condenser (refrigeration)**:** Is a heat exchanger in which a vapour, such as steam, condenses to the saturated liquid state as the result of heat transfer from the vapour to a cooling medium such as a lake, a river, or the atmosphere.

Conduction: Is the transfer of energy from the more energetic particles of a substance to the adjacent less energetic ones as a result of interaction between particles.

Conductor: A metallic unit, which is a carrier of electrical current.

Conduit: 1. Any metallic or other pipe, flexible or rigid through which electric cables are carried. 2. Any open or closed channel intended for the conveyance of water.

Cone clutch: A form of friction clutch in which the power necessary for driving is effected by the friction of smooth, turned conical surfaces, often faced with leather.

Cone key: In machinery, a form of key used for retaining a wheel or pulley in place, when the hole in the wheel is larger than the portion of shaft upon which it is keyed. The wheel is bored slightly conical and a conical ring turned to fit in the bored hole and to embrace the shaft. It is then slotted into three parts, forming three separate keys. The wheel is thus maintained concentric with the shaft, and will pass over a larger to a smaller section without the necessity for splitting it.

Cone nozzles: A hydraulic spray nozzle, which produces a circular spray, pattern.

Cone pulley: A stepped pulley, one having two or more faces of different diameters; used in pairs, the large end of one being opposite the small end of the other so that a shifting of the belt will give a change of speed. *See* Fig. 16C.

Conical spray: Spray with a conical shape.

Connecting rod: A rod or arm, with bearings at both ends, which makes the connection between crankshaft and pistons.

Connecting rod bearing: The bearing at the large end of connecting rod, where it fits on the crankshaft, is halved or split and is usually of Babbitt metal. The bearing at the small end is of the bronze bushing type.

Conservation of energy principle: Energy can be neither created nor destroyed; it can only change forms. The net change (increase or decrease) in the total energy of the system during a process is equal to the difference between the total energy entering and the total energy leaving the system during that process.

Constant mesh gears: Gears, which remain constantly in, mesh as contrasted with gears in the transmissions, which are shifted out of or into mesh.

Constant volume injection (CVI) system: A type of port injection fuel delivery system in which air is injected separately at the beginning of the intake stroke to dilute the hot residual gases and cool any hot spots.

Continental cambridge roller: Roller comprising discs of unequal diameter of alternating smooth and notched surface on the same shaft.

Contour furrows: Furrows ploughed on the contour on pasture to prevent soil loss and allow water to penetrate into the soil. Also furrows are laid out on the contour for irrigation purposes.

Contour ploughing: The method of ploughing in which the soil is broken and turned along contours.

Control surface: The boundary of a control volume, and it can be real or imaginary.

Control volume, or open system: Any arbitrary region in space through which mass and energy can pass across the boundary. Most control volumes have fixed boundaries and thus do not involve any moving boundaries. A control volume may also involve heat and work interactions just as a closed system, in addition to mass interaction.

Convection: is the transfer of energy between a solid surface and the adjacent fluid that is in motion, and it involves the combined effects of conduction and fluid motion.

Conventional ply tyre: A tyre in which the cords of the body plies run diagonally from bead to bead.

Conventional tillage: The combined primary and secondary tillage operations normally performed in preparing seedbed.

Converging diverging nozzles: Are the ducts in which the flow area first decreases and then increases in the direction of the flow.

Conveyor: Device such as an endless belt, or chain with buckets attached for transferring material.

Conveyor fed chaff cutter: A chaff cutter in which the feeding of the fodder crop is done through a conveyor.

Conveyor platform (thresher)**:** An assembly for carrying the cut crop to the threshing unit or to the feeder conveyer.

Coolant: A liquid used to transfer heat to or from engine components. It consists of either pure deionized water, a mixture of de-ionized water with pure ethylene glycol, or standard antifreeze depending on the circuit.

Cooling system: The group of devices, which prevent the overheating of the engine by rapidly carrying off the heat generated by combustion. Radiator fans, pump and water jackets are parts of a water-cooling system. Fins placed on the piston cylinder serves as air cooling system.

Corrugated roller: Roller consisting of one or more cylinders of which the surface has corrugations in a concentric form or parallel to the axis of the cylinder. *See* Fig. 17C.

Cotter: A tapered rod or pin, generally flat in section, used for wedging the ends of rods.

Cotter pin: Usually a form of split pin, which is inserted into a hole near the end of bolt to prevent a nut from working loose.

Cotton stalk puller: A device to pull the cotton stalks from the field.

Coulter: A device to cut the soil vertically from the land ahead of the plough bottom. Two types of coulters are available as: 1. Rolling type disc coulter and 2. Sliding type knife coulter. *See* Fig. 18C.

Countershaft: The shaft in the lower part of the transmission which carries the gear cluster driven by the clutch shaft pinion gear and which, in turn, drives the sliding gears of the transmission, also called lay, idler, or jack shaft.

Countershaft gear: The cluster of gears on the countershaft within the lower portion of the transmission housing.

Couple: Two equal and opposite forces producing rotational movement.

Coupling: An attachment whereby one piece of mechanism constrains another part to follow the movements of the first. It may be an absolutely rigid connection as a flanged coupling to unit sections of shafting together or flexible within limits as a universal joint or a diaphragm coupling.

Covering device (seed drill) **:** A device to refill a furrow after seeds have been placed in it.

Covering discs (potato planter) **:** Paired concave discs used to throw a shallow ridge of soil over plant potatoes.

Crank: A lever, which rotates about the axis of a shaft.

Crankcase: The lower part of the engine in which the crankshaft and many other parts of the engine operates, includes the lower section of the cylinder block and the oil pan. It acts as a reservoir for the supply of lubricating oil.

Crankcase dilution: Unburned fuel passing out from the piston rings and dripping into the crankcase, where it dilutes or 'thins' the engine lubricating oil. This is called as crankcase dilution.

Crankcase ventilation: The circulation of air through the crankcase of a running engine to remove water and gasoline vapors to prevent oil dilution, contamination, sludge formation and pressure buildup.

Crankshaft: The main drive shaft of an engine, which takes reciprocating motion and converts it to rotary motion. *See* Fig. 4C.

Crankshaft counter balance: A source of weights attached to or forged integrally with the crankshaft to offset the reciprocating weight of each piston rod.

Creep: A movement of one element relative to the other with which it is in contact e.g. slipping motion between a belt and the pulley upon which it revolves.

Critical point (thermodynamics)**:** The point at which the saturated liquid and saturated vapour states are identical.

Crossed belt: A driving belt, which has a twist between the driving and the driven pulleys causing a reversal of direction.

Cryogenic temperatures: Temperatures below -100 °F (-200^0 K; -73 °C).

Cubic centimeter: A volume of space equal to a cube that is one centimeter on each side. Abbreviated as cc or cm^3, and equivalent to a milliliter (ml) in capacity.

Culti packer: Tandem combination in the same frame either of two cambridge rollers (deep and wide corrugations type) offset by half a disc or of a cambridge and cross kill rollers.

Cultivator: An implement for inter-cultivation with laterally adjustable tines or discs to work between crop rows. This can be used for seedbed preparation and for sowing with seeding attachment. The tine may have provision for vertical adjustments also. *See* Fig. 19C.

Cultivator point: A replaceable soil working part attached to the lower end of the cultivator tines as shovel and sweep.

Cultivator tine: A member connecting cultivator point to mainframe. *See* Fig. 19C.

Cup feed mechanism: A feed mechanism consisting of cups or spoons on the periphery of a vertical rotating discs which picks up seeds from hopper and delivers them into a seed tube. *See* Fig. 20C.

Cup-feed seed drill: A seed drill, the metering mechanism of which consists of discs fitted with cups on their periphery. The seed rate is controlled either by altering the size of the cups or by changing the speed of rotation of the discs.

Curing: Natural or artificial ageing of the agricultural produce brought about usually by some (dry or wet) heat treatment.

Cut off device (plant protection appliances)**:** A hand operated mechanism situated between the delivery hose and the spray lance for controlling the flow of the liquid.

Cut off mechanism (seed or fertilizer metering)**:** A device, which cuts off or brushes out excess seeds from the cells of feed mechanism.

Cutter bar: The assembly comprising finger bar, fingers, knife guides, wearing plate, outer shoe and main shoe, i.e., the non-reciprocating part of the cutting mechanism. It is used for cutting grasses and forage. *See* Fig. 21C.

Cutter bar effective width: The distance between the point at which tips of the knife sections meet the last effective shearing edges of the guards or shoes at the extremities of the cutter bar expressed in millimeters.

Cutter bar height: The height of the forward tip of any knife section above the plane on which the combine in standing expressed in millimeter. Where the height is adjustable, maximum and minimum value should be measured.

Cutter bar knife: A reciprocating part (steel plate) of cutter bar, having triangular shape with two cutting edges. *See* Fig. 21C.

Cutter bar ledger plate: A hardened metal inserted in a guard (finger) over which knife section move to give a scissor like cutting action. *See* Fig. 21C.

Cutter bar shoe: Are provided for regulating the height of crop above the ground.

Cutter bar wearing plate: A hardened steel plate attached to the finger bar to form a bearing surface for the back of the knife. *See* Fig. 21C.

Cutter bar working width: The distance expressed in millimeters between two vertical planes passing through the point of the outermost divider and parallel to the centerline of the cutter bar. If adjustable dividers are used, the maximum and minimum dimension should be stated.

Cutter loader: A machine to cut crops and deliver them to vehicle.

Cutting angle: The angle measured between the cutting face of the tool and the surface of the material on which the tool operates.

Cutting edge: The front edge of share, which makes horizontal cut in the soil. *See* Fig. 11C.

Cutting means (blade) **:** The mechanism used to provide the cutting action.

Cutting means enclosure: The part or assembly, which provide the protective means around the cutting means.

Cutting oils: Any of the heavy oils or combination of oils used as a metal lubricant in machining operations. This term does not include the watery solutions used as coolants.

Cutting positions: Any designated height adjustment of the cutting element where it remains operative.

Cutting tools: A general term but referring to edged tools used in machines more than the tools held by hand.

Cutting width: The width of cut measured across the cutting element at right angles to the direction of travel and calculated from the dimensions of the cutting element or the diameter of the blade tip circle.

Cyclone separator: It is a separating device in which strong centrifugal force acting radially is used in place of relatively weak gravitational force acting vertically downwards.

Cylinder (engine) **:** A round hole having some depth, bored to receive a piston. It is also referred as a 'bore'. *Refer* Fig. 2A.

Cylinder (threshing)**:** A balanced rotating assembly comprising of beater bars or spikes on its periphery and their support for threshing the crop. *See* Fig. 22C.

Cylinder and concave clearance: The gap between the tip of the cylinder to the inner surface of the concave expressed in millimeter.

Cylinder block (engine)**:** The main body of the engine, which is bored to receive the pistons. The cylinder block and crankcase are frequently cast as one piece.

Cylinder bore (engine)**:** The internal diameter of an engine cylinder.

Cylinder diameter (threshing)**:** The diameter of the circle generated at the outermost point of the cylindrical threshing element expressed in millimeter.

Cylinder head (engine)**:** A detachable portion of an engine fastened securely to the cylinder block, which is a portion of a combustion chamber.

Cylinder length (threshing)**:** The distance between the outermost points of the cylinder threshing elements. *See* Fig. 22C.

Cylinder liner (engine)**:** A sleeve or tube interposed between the piston and the cylinder wall, or cylinder block to provide a readily renewable wearing surface for the cylinder.

Cylinder mower: A mower with rotating helical blades arranged in horizontal cylindrical form.

Cylinder type chaff cutter: A chaff cutter, the cutting mechanism of which consists of a rotating cutting cylinder.

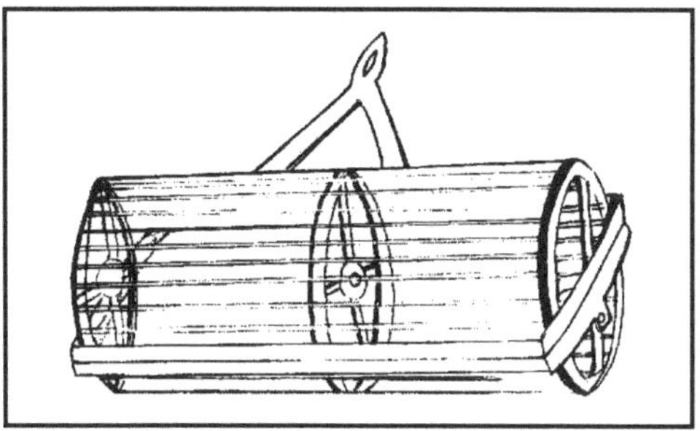

Fig. 1C Cage roller

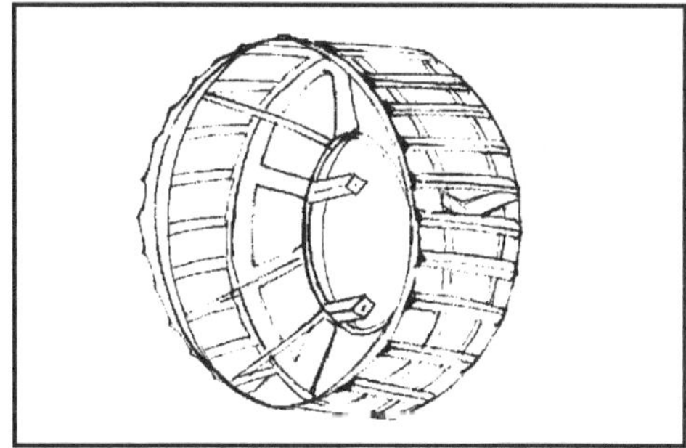

Fig. 2C Cage wheel

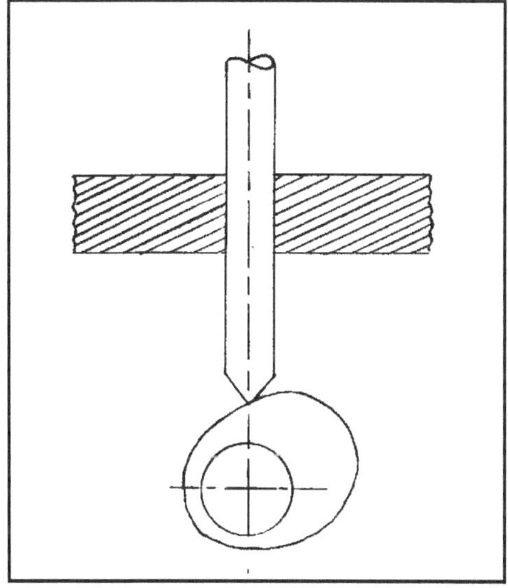

Fig. 3C Cam and follower

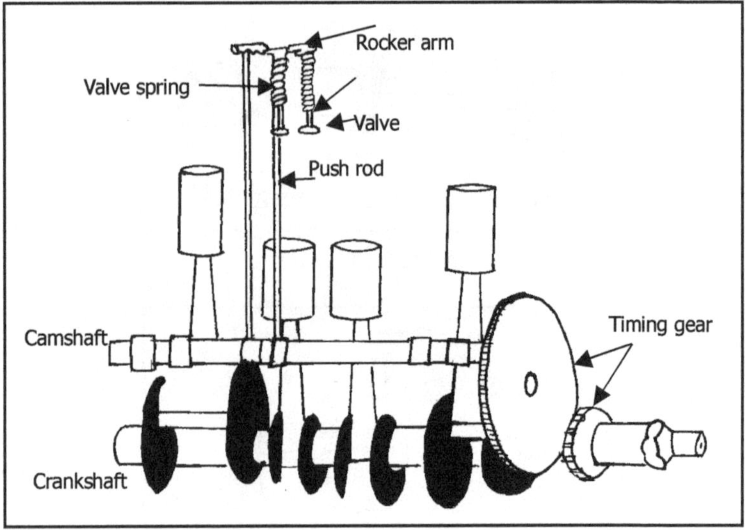

Fig. 4C Components of IC engine

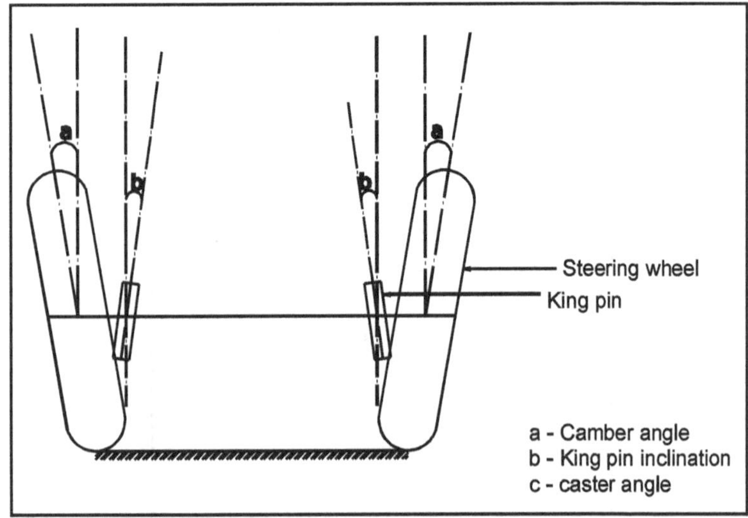

Fig. 5C Camber angle

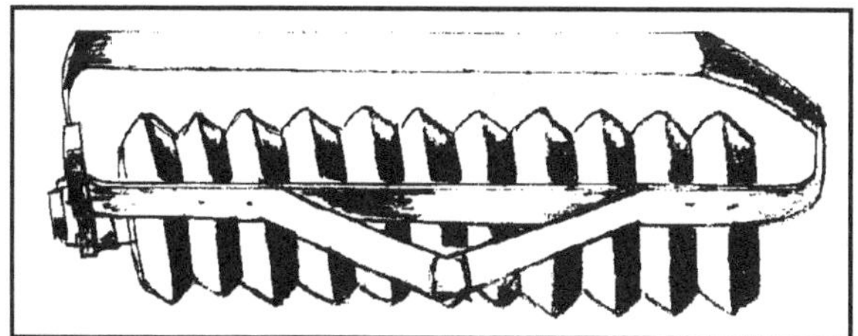

Fig. 6C Cambridge roller

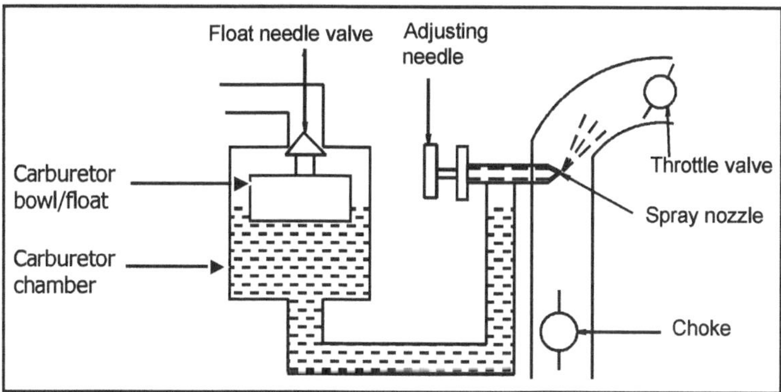

Fig. 7C Carburetor

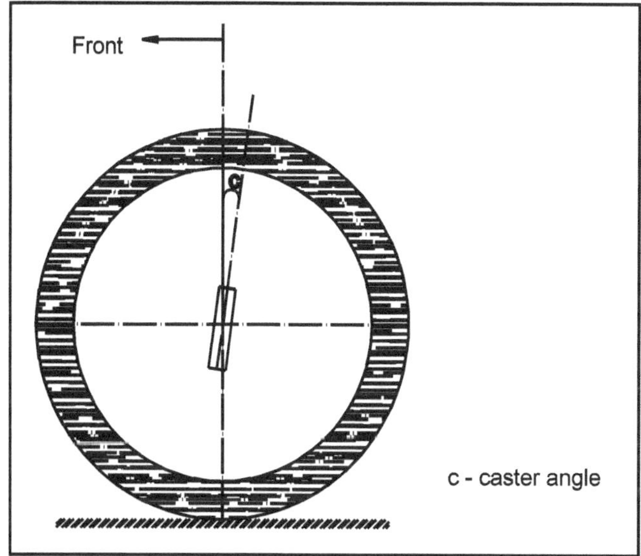

Fig. 8C Caster angle

Fig. 9C Center punch

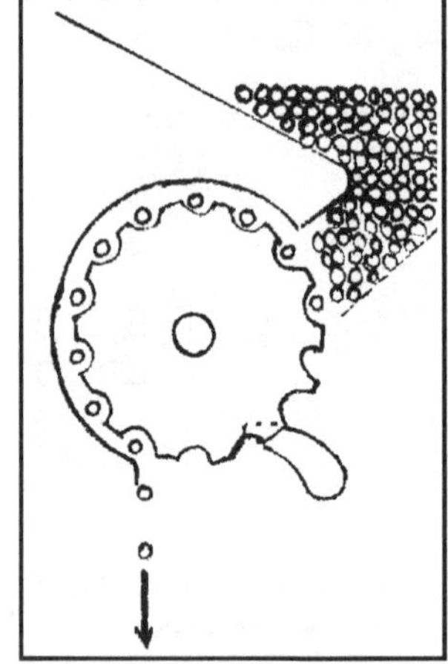

Fig. 10C Cell feed mechanism

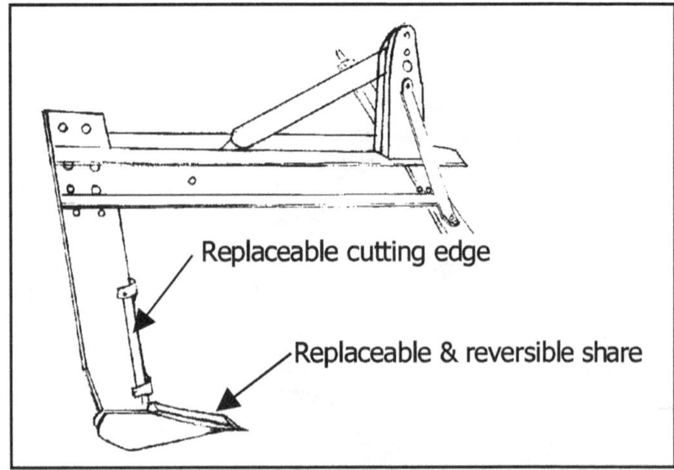

Fig. 11C Chisel plough

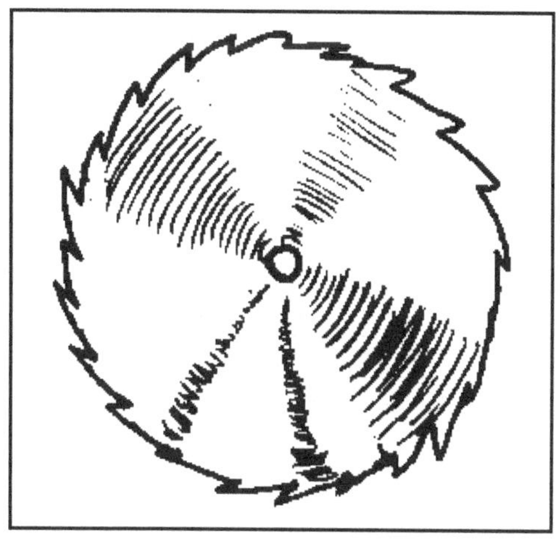

Fig. 12C Circular saw

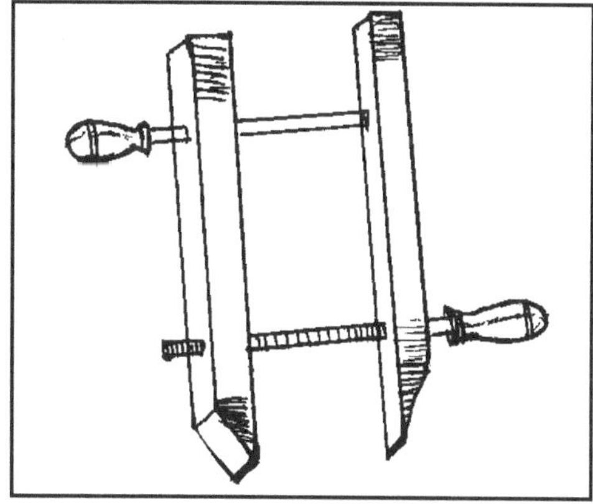

Fig. 13C Clamp

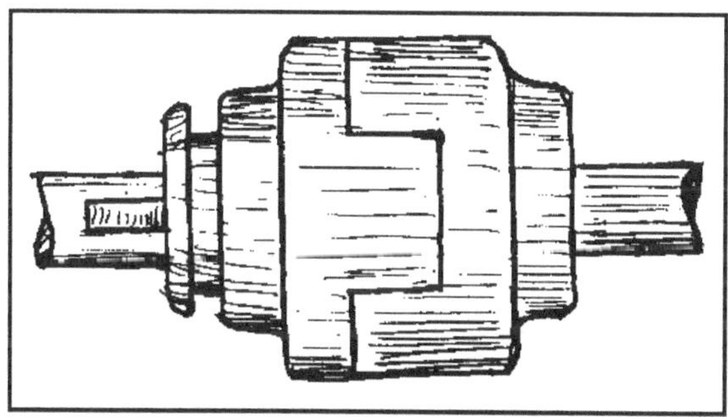

Fig. 14C Claw coupling

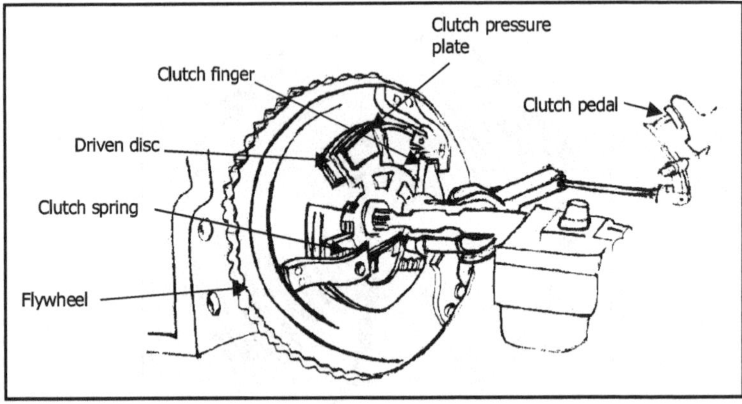

Fig. 15C Clutch

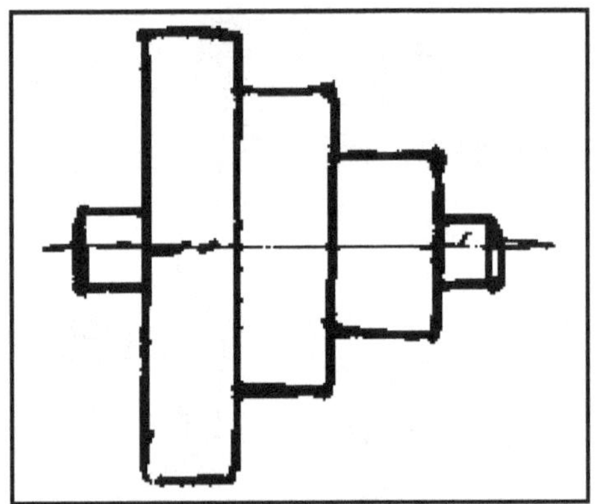

Fig. 16C Cone pulley

Fig. 17C Corrugated roller

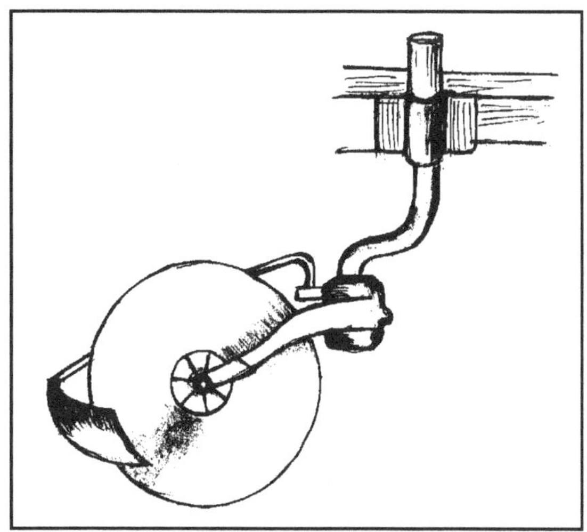

Fig. 18C Coulter

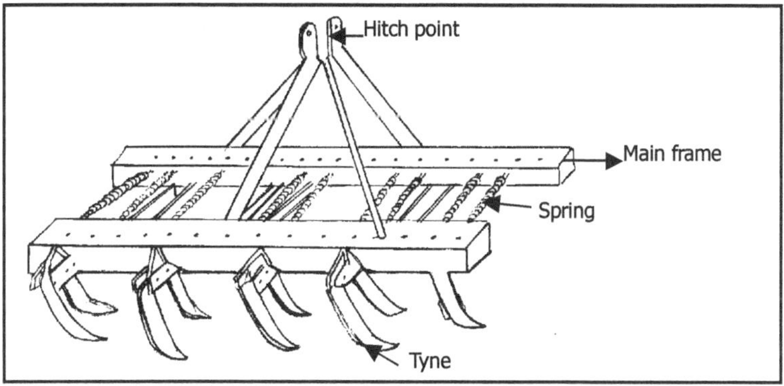

Fig. 19C Cultivator (Tractor drawn spring type)

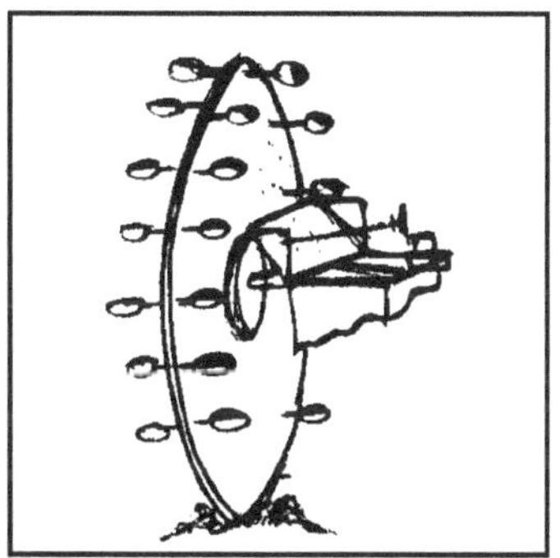

Fig. 20C Cup feed mechanism

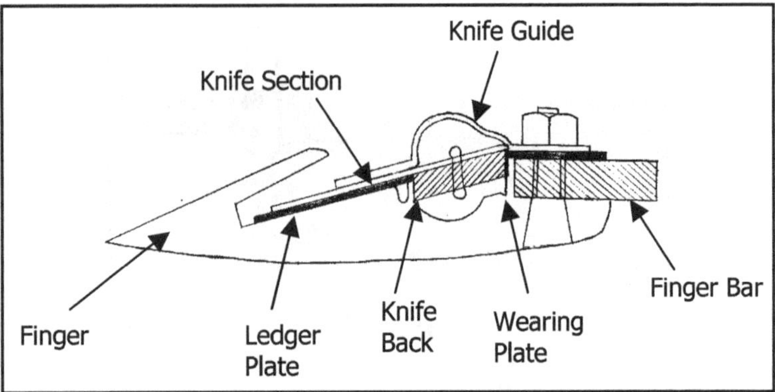

Fig. 21C Cutter bar

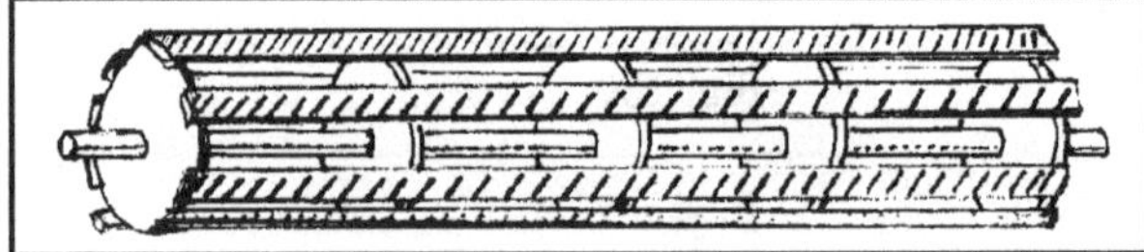

Fig. 22C Cylinder (threshing)

Dah: A metal with handle used for cutting trees and cleaning jungle growth. *See* Fig.1D.

Dead axle (tractor trolley)**:** An axle beam, which does not drive the road wheels. Usually refers to a rear axle which does not have any rotating parts within it, but on which the rear wheels are mounted with antifriction bearing.

Dead centre: 1. The centre, which is fitted into the tailstock of a lathe and does not rotate with the work. 2. The extreme top or bottom position of the crankshaft throw at which the piston is not moving in either direction.

Dead furrow: An open trench left in between two adjacent strips of land after finishing the ploughing operation. *See* Fig.1B

Dead weight: The weight of a vehicle or carrier itself as distinguished from carried or live load.

Debarker: A machine used for removing the bark from the tree.

Debarking: Operations of removing bark from trees or parts of the tree.

Debranching: Operations of removing branches from trees or parts of the trees.

Deceleration: A decrease in velocity or speed with time, allowing an engine to coast to idle speed from a higher speed with the accelerator at or near the idle position.

Decelerometer: A device used to determine the rate of deceleration.

Decompression valve: Device for lowering the compression in the cylinder to aid starting.

Dedendum (gear): The portion of a gear tooth lying between the pitch and the root lines. *Refer* Fig.9G.

Deep percolation: Downward movement of the water below the plant root zone. Deep percolated water can not be used by the plants.

Deep tillage: A primary tillage operation, which manipulates soil to a greater depth than normal ploughing. It may be accomplished with a heavy-duty mould board or disc plough, which inverts the soil, or with a chisel plough or subsoiler, which shatters soil.

Deflector nozzle: A hydraulic spray nozzle with a deflector producing a flat sheet of spray of full circular pattern or narrower pattern of 25 to 180°.

Deformation: The alteration in form of a structure when subjected to the action of load.

Deformation curve: A curve showing the relationship between the stress or load on a specimen and the resulted strain. Also known as stress strain curve.

Defueling: To release the fuel from a container or vessel.

Degree of polish (rice)**:** The quantity of bran removed from brown rice during polishing expressed in terms of percentage of brown rice.

Dehumidifier: Equipment designed to reduce the amount of water form vapour in the ambient atmosphere. *See* Fig.2D.

Dehumidifying: The process of removing moisture from atmospheric air.

Dehydration: The rapid removal of moisture to a very low level by applying external heat.

De-ionized water: Water, with its ions removed to make it nonconductive. Also known as distilled water.

De-ionizing filter: A filter that removes ions from a liquid.

Delimber: Machine, designed to remove the branches from trees.

Delinter: Machine, used to remove residual lint (fuzz) from cottonseed.

Density: The amount of mass contained per unit volume. In case of gas, density values only have meaning at a specified temperature and pressure since both of these parameters affect the compactness of the molecular arrangement. The density of a gas is called as vapour density, and the density of a liquid is called as liquid density.

Depreciation: Loss of value due to physical deterioration.

Depth gauge: A gauge, being used in wood and metal works for testing the depth of holes and recessed portions.

Depth micrometer: A precision gauge, used to determine the depth of holes, slots, counter bores or the distance from one surface to a lower level etc.

Depth of cut: The maximum depth of the penetration of the tool measured with reference to the initial soil surface.

Destoner: A mechanical device, which separates stones from the grain.

Detergent: Soap like compound, used in engine oil to remove engine deposits and hold them suspended in the oil.

Detonation: The very rapid burning of vapor resulting in a self-sustaining shock wave, the pressure behind which is several atmospheres. Detonation waves travel at speeds exceeding the speed of sound in air. In an internal combustion engine, detonation is commonly referred to as spark knock or ping. In the combustion chamber, an uncontrolled second flame front (after the spark occurs at the spark plug) with spontaneous combustion of the remaining compressed air: fuel mixture, resulting in a pinging noise.

Device: A mechanical aid or contrivance, which serves to promote the better performance of a job.

Dew-point temperature: The temperature at which condensation begins when the air is cooled at constant pressure.

Diagnosis (mechanical): A procedure followed in locating the cause of a failure.

Diagnostic code: A code obtained from a computerized diagnostic tool.

Diagram: An outline drawing or graphical representation.

Dial: A graduated plate, usually circular or oval on which a needle or pointer indicates the reading.

Dial gauge: A dial having an index hand actuated by a bent spring, through which the amount of pressure or vacuum is indicated. All gauges of this kind are made on bourdon's spring principle.

Diamond chisel: A chisel having a V shaped or diamond shaped cutting edge.

Diaphragm: A thin dividing sheet or partition that separates an area into compartments; used in fuel pumps, modulator valves, vacuum advance units and other control devices.

Diaphragm pump: A volumetric pump in which the flow of the liquid is achieved by the deformation of diaphragm.

Dibbler: A device used for dibbling. *See* Fig.3D.

Dibbling: The process of placing seeds in the holes made in a seedbed and covering them.

Die (mechanical): Are sunk, i.e. the cavities are formed by milling, electric discharge machining, or sometimes by ceramic casting processes.

Die-casting: The art of rapidly producing accurately dimensioned parts by forging molten metal under pressure into split metal dies, which resemble a common type of permanent mould.

Dielectric: A nonconductive, insulating material.

Diesel cycle: The ideal cycle for compression ignition reciprocating engines, first proposed by Rudolf Diesel in the 1890s. Diesel cycle consists of four internally reversible processes as: 1-2 Isentropic compression; 2-3 Constant pressure heat addition; 3-4 Isentropic expansion; 4-1 Constant volume heat rejection. *See* Fig.4D.

Diesel engine: Named after its designer, Dr. Rudolph Diesel. This is an internal combustion engine of high efficiency, ignites fuel in the cylinder from the heat generated by compression. The fuel is an 'oil' rather than gasoline and no spark plug or carburetor is required.

Diesel fuel: Diesel fuel is the most common fuel for heavy-duty compression ignition engines. It is a standard of comparison for other fuels.

Diesel volume equivalent (DVE): The volume of hydrogen in cubic metre to a litre of diesel on an energy equivalent basis.

Differential: A device to enable the two wheels, driven from a single shaft to rotate at different speeds. While turning a vehicle on a curved path, the inner wheel has to turn slower and the outer has to turn relatively faster. The differential unit fulfills this condition. *See* Fig.5D.

Differential lock: A device when engaged would permit two driven wheels on the separate shafts to rotate at same speed with common drive shaft. *See* Fig.6D.

Diffuser: A device that increases the pressure of a fluid by decreasing the fluid velocity.

Diffusivity: The ability of a gas to diffuse in air.

Digging fork: Fork with long handle used for digging. *See* Fig.7D.

Dimension: Physical characterizations of a quantity.

Dimensioning: Indicating on a drawing the sizes of various parts.

Dimmer: A device for reducing the glare of headlights. The usual practice is to raise and lower the headlight beam by means of a switch. Or a rheostat placed in series with a lamp to control its brilliancy.

Dipstick: An engine fluid level indicator.

Direct current (DC): A current, that flow through a circuit in one direction only. Direct current voltage is designated VDC.

Direct current arc welding: An arc-welding process where in the power supply at arc is direct current.

Direct injection system: A sophisticated system that forms the fuel-air mixture inside the combustion cylinder after the air intake valve has closed.

Direct fired dryer: A dryer in which the product of combustion comes into direct contact with the product being dried.

Directional nozzle: A type of nozzle, which enables the direction of spray to be altered in relation to the angle of the supply tube or pipe.

Directional signal: An electrical device used to indicate the direction in which the driver of an automobile expects to turn.

Disc: A circular, concave and revolving steel plate used for cutting and inverting the soil. *See* Fig.9D.

Disc angle: The angle at which the plane of cutting edge of the disc is inclined to direction of travel. Disc angle is in the range of 42^0 to 47^0. Increasing the disc angle improves disc penetration. *See* Fig.8D.

Disc coulter: A circular flat or concave revolving steel disc fitted to the beam through a shank and yoke. *Refer* Fig. 18C.

Disc harrow: It is a harrow which performs the harrowing operations by means of a set of rotating steel discs, each set being mounted on a common shaft. Tractor drawn disc harrows are divided in to two classes as 1. Single action disc harrow and 2. Double action disc harrow. Refer single action disc harrow and double action disc harrow.

Disc spacing: The transverse distance between the edges of two discs separated by a spool.

Discharge chute: An extension of the discharge opening generally used to control the discharge of material from the cutting means.

Discharge height of unloader (combine): The vertical distance from a plane on which the combine is standing to the lowermost point under discharge opening with the unloader in operating position expressed in millimeters.

Disc-selector drill: A spacing drill, the metering mechanism of which consists of a disc with notches mounted either vertically or inclined. Seed spacing is usually controlled by varying the speed at which the disc rotates or by selecting discs with different notch spacing.

Disengaging clutch: A clutch for the throwing of a line of shafting or a train of wheels into and out of gear.

Disk clutch: A disk is held against a plate by springs with such force that there is no slipping during the transmission of power through the clutch. Clutches are engaged and disengaged by means of the clutch pedal. *Refer* Fig.15C.

Disk harrow: A harrow which performs the harrowing operation by means of set (or number of sets) of rotating steel discs, each set being mounted on a common shaft. *See* Fig.9D & 10D.

Disk plough: A plough the working parts of which are concave or truncated cone discs, which act simultaneously as mould board and share. *See* Fig.11D.

Displacement volume: The volume displaced by the piston as it moves between top dead centre and bottom dead centre.

Dissipation of energy: The use of energy without a commensurate work return (wastage of energy).

Distance ratio: The ratio of the distance moved by the effort or input of a machine in a specified time to the distance moved by the load or output.

Distilled water: Water, which has been changed to a gas (steam) by boiling, then condensed (change back to liquid) by cooling. The process is used to eliminate harmful minerals or other foreign materials.

Distortion: A change from the original shape.

Distributing mechanism: A device, which helps in spreading.

Distribution efficiency: Measure of the uniformity of distribution.

Distributor (liquid fertilizer): A machine, for applying mineral fertilizer in liquid form, which either spreads the fertilizer on the surface of the ground or injects it into the soil.

Distributor (gaseous fertilizer): A machine, for applying mineral fertilizer in gaseous form, which injects fertilizer under pressure into the ground.

Distributor (spark ignition system): The part of the spark ignition system for interrupting the low voltage primary current and distributing the resulting high voltage current to the engine cylinder in proper sequence and time. The components of timer distributor are housing, drive shaft, cam, breaker point, rotor and cap. Distributor is also called as Ignition timer distributor. *Refer* Fig.6B.

Ditcher: An implement used for making ditches.

Divider: An attachment to a shoe mainly for dealing with tangled crops.

Dog: The carrier of a lathe. One of the jaws of a chuck.

Dog clutch: Clutch having square jaws, mostly used in power tillers.

Dolly: A tool held against heads of rivets while riveting.

Double action disc harrow: A disc harrow consisting of two or more gangs in which a set of one or more gangs follows behind the set of the other one or more, arranged in such a way that the front and back gangs throw the soil in opposite directions. It is of two type as 1. Tandem and 2. Offset. Refer tandem disc harrow and offset disc harrow. *See* Fig.10D.

Double disc furrow opener: A furrow opener consisting of two flat discs, set at an angle to each other.

Double end bolt: Solid shaft having threads on both the ends. *See* Fig.12D.

Double plate clutch: A clutch of the friction type with two driven plates.

Dowel pin: A pin inserted in matching holes in two parts to maintain these parts in fixed relation one to the other.

Down draft carburetor: A type of carburetor in which the fuel-air mixture flows downwards to the engine.

Draft: The horizontal component of the pull.

Drafting: The art of drawing.

Drafts man: One who prepares drawings, or portraits graphically; usually applied to one who uses mechanical aid or instrument in his work.

Drag link: The rod, which connects the left steering- knuckle arm with the steering gear arm. There are two cups and a spring at each end to engage the ball ends of these arms.

Dragging brakes: Brakes, which because of improper adjustment drag their linings on the breaking surface of the drum, thus causing rapid lining wear, over heating of the engine, and other troubles.

Draining and tapping knife: A blade attached to a wooden handle and is designed to have a V shaped cutting edge to make narrow channels in the bark of rubber trees to open the latex cells and to facilitate easy tapping and draining of latex from the tree.

Draw: (1) To portray by a system of lines. (2) To bring the temper of steel from extreme hardness to the desired hardness or temper. (3) To form by a stretching or distorting process; as to draw out wire.

Drawbar: A member fitted to a tractor, through which the pulling force to an implement or trailer is applied to an implement. It may be swinging type, fixed type or adjustable type.

Drawbar dynamometer: Used for measurement of drawbar horsepower of the engine. Spring dynamometer, hydraulic dynamometer and strain gauge dynamometer are the types of drawbar dynamometer.

Drawbar horse power (DBHP): Power obtained at drawbar with the governor control in the position recommended by the tractor manufacturer for drawbar work and the tractor moving on the horizontal surface with the drawbar pull applied horizontally.

Drawbar pull: It is the pull exerted by a tractor at its drawbar.

Drill: A tool for boring holes in metal or wood.

Drill chuck: A chuck made especially for holding drills. *See* Fig.13D.

Drill gauge: A flat steel plate drilled with holes of different sizes and properly marked so that the size of a drill may be easily determined by fitting it to plate.

Drilling: A process of making holes in metal or wood.

Drive shaft: The propeller shaft, which transmits power from the forward universal to the rear universal or pinion shaft.

Driven plate: The part of the clutch, which receives the engine power from the flywheel and the pressure plate and transmits it to the clutch shaft and transmission, is known as clutch plate. *See* Fig.15C.

Driving pinion: The input gear in the differential of an automobile.

Driving plate:. A part of the clutch which is bolted to the flywheel to receive power from the engine and transmit it by means of pressure and antifriction to the driven plate, clutch shaft and transmission. Also called as pressure plate. *See* Fig.15C.

Drop centre rim: A solid piece tyre rim having a deep channel or well rolled into its middle. In mounting or dismounting a tyre, the beads are dropped into the well, thus allowing them to clear the flanges of the rim.

Drop leg (spray boom)**:** An auxiliary vertical spray boom fixed below a main horizontal spray boom.

Droplet size: The mean diameter of the droplet expressed in micrometers.

Dry air: Air that contains no water vapour.

Dry cell: The small commercial cell used as a source of current in flashlights, doorbells, etc. It produces electrical current by chemical action; called 'dry' because the chemical are in the form of a moist paste instead of liquid as in the storage cell.

Dry clutch: A friction clutch designed to operate without the application of any lubricant or fluid.

Dry gas: A gas that does not contain water vapor.

Dry mass of power tiller: The mass of power tiller fitted with all component necessary for its operation but without water, fuel and oil.

Dry-bulb temperature: The ordinary temperature of atmospheric air.

Dry type air cleaner: The filtering element in this is a type of felt. The large surface area of filtering element reduces incoming air speed and hence foreign particles from the air are removed by depositing at the surface.

Dryer: A unit, which provides the condition for reducing moisture generally by, forced ventilation with or without addition of heat.

Drying: The reduction of moisture in a product usually to some predetermined moisture content.

Drying air: The air being passed through the grain being dried.

Drying air temperature: The temperature of the air entering the product being dried.

Drying conveyor: A mechanical linkage or assembly to carry the product being dried through the drying chamber.

Dual cycle: Is the ideal cycle which models the combustion process in both gasoline and diesel engines as a combination of two heat-transfer processes, one at constant volume and the other at constant pressure.

Duel fuel engine: Internal combustion engines that can operate on either of two fuel, such as natural gas or gasoline.

Ductility: The property of materials, which permits permanent deformation without failure after the elastic limit, has passed. Drawing metal into a wire is a common example of ductility.

Duplex carburetor: A carburetor which has a double ventury for mixture distribution.

Duplex pump: A reciprocating pump with two parallel pumping cylinders.

Dust nozzle: A device, by which pesticide in the form of dust is directed in airflow.

Duster: An appliance used for dusting.

Dusting: Application of plant pesticide in the dust form.

Dynamic traction ratio: Ratio of drawbar pull to actual force on traction device normal to traction surface while pulling.

Dynamic weight (traction device)**:** Total force exerted by the traction device normal to the supporting surface under operating conditions.

Dynamics: That branch of mechanics, which treats of the laws of force that procedure motion in bodies.

Dynamite: An explosive composed of an absorbent saturated with nitroglycerin.

Dynamo: It is a device on tractor or any vehicle to generate voltage for charging battery, which in turn used to supply current for ignition, light etc. It consists of armature, which is rotated between permanent magnetic cores. Voltage is generated due to rotation of armature, which is carried to the commutator and from commutator through brushes to the battery. *See* Fig.14D.

Dynamometer: A test unit for measuring the actual power produced by an engine. Dynamometers are commonly classified as 1. Brake dynamometer and 2. Drawbar dynamometer. Refer brake dynamometer and drawbar dynamometer.

—□—□—

Fig. 1D Dah

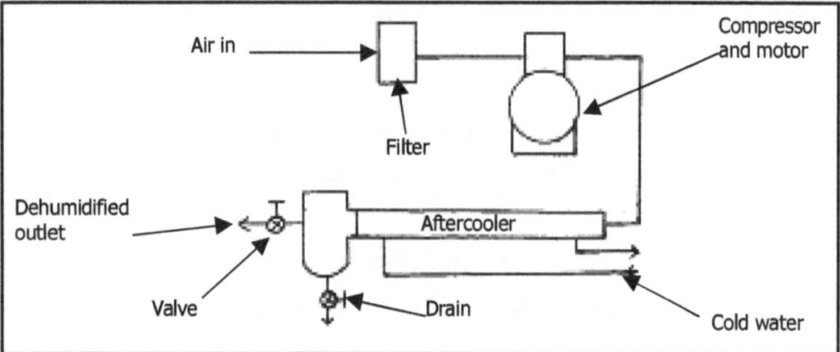

Fig. 2D Dehumidifier

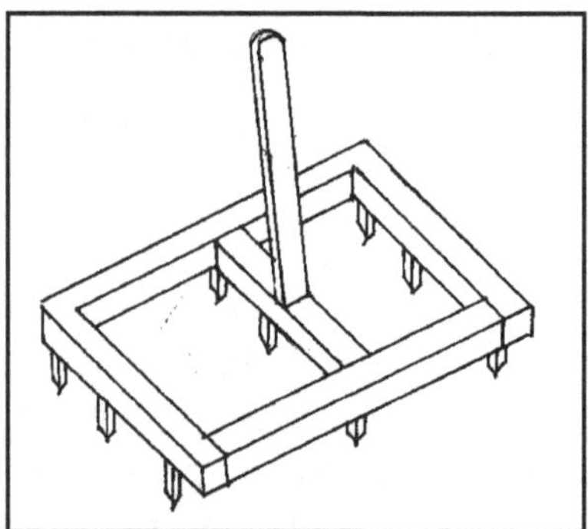

Fig. 3D Dibbler

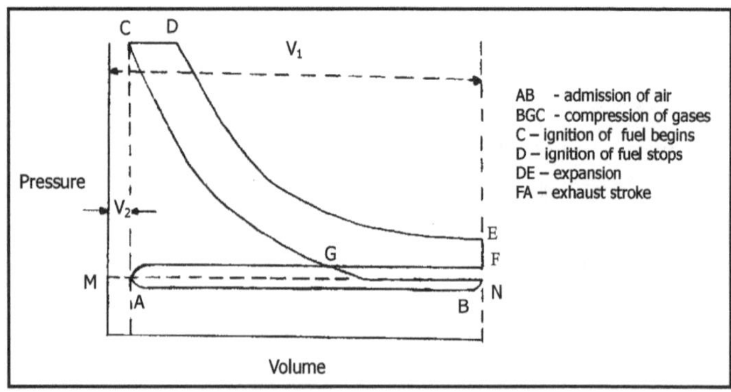

Fig. 4D Diesel cycle

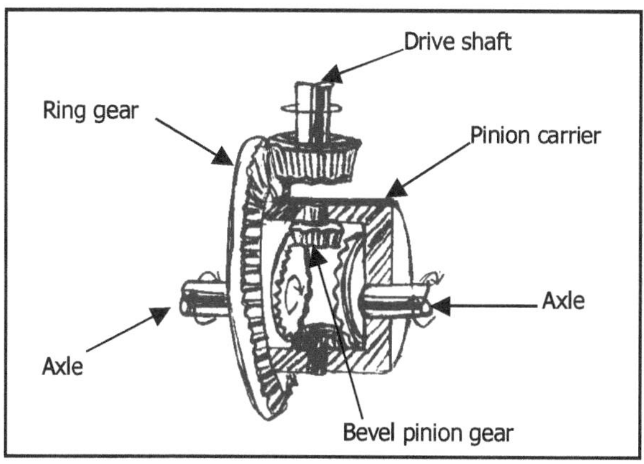

Fig. 5D Differential

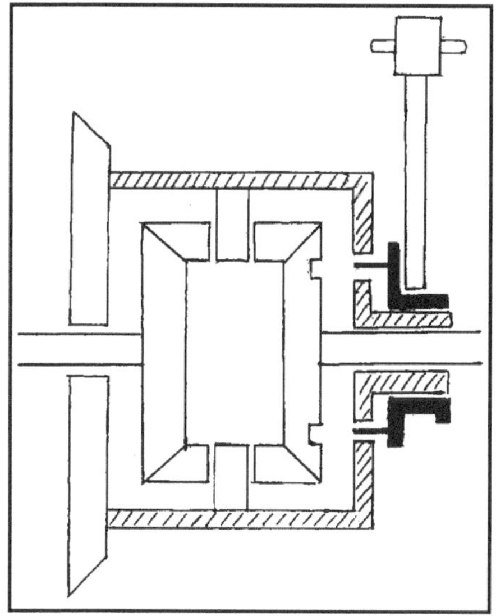

Fig. 6D Differential lock

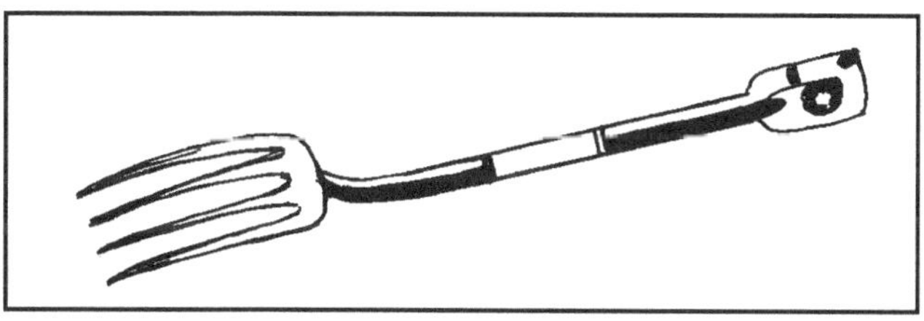

Fig. 7D Digging fork

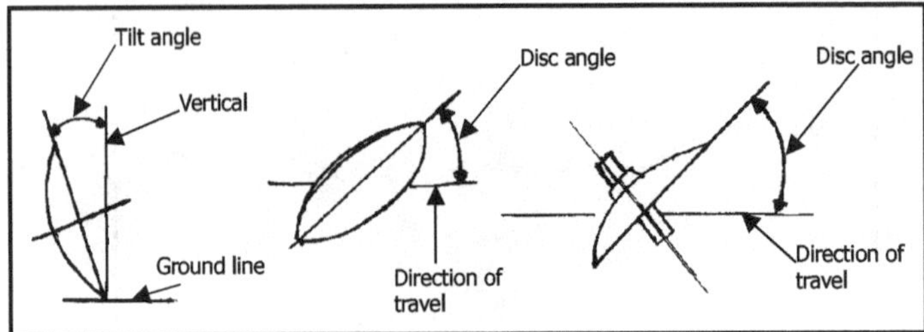

Fig. 8D Disc angle and tilt angle

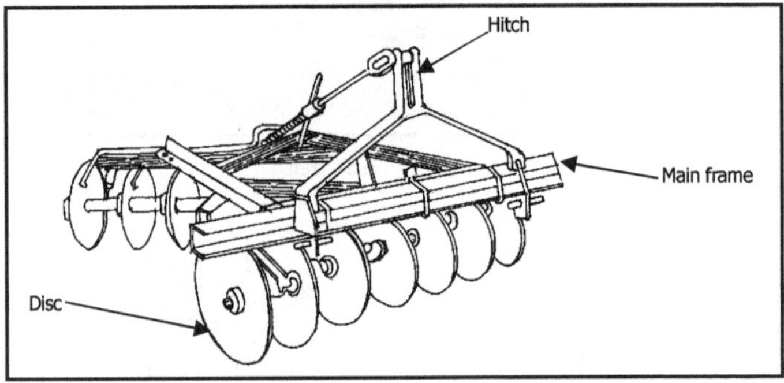

Fig. 9D Disc harrow (Offset type)

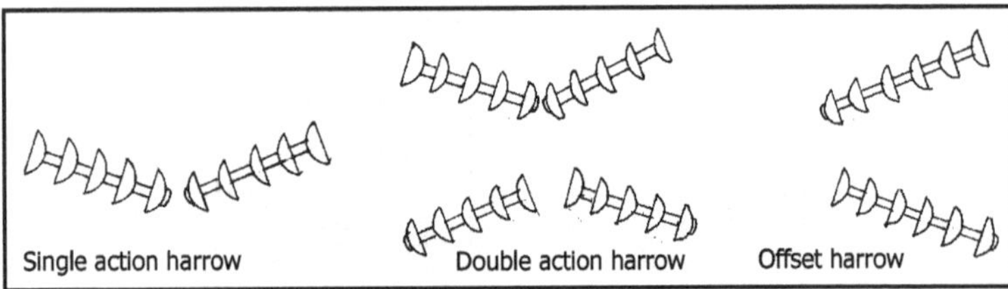

Fig. 10D Disc harrows (arrangement of gangs)

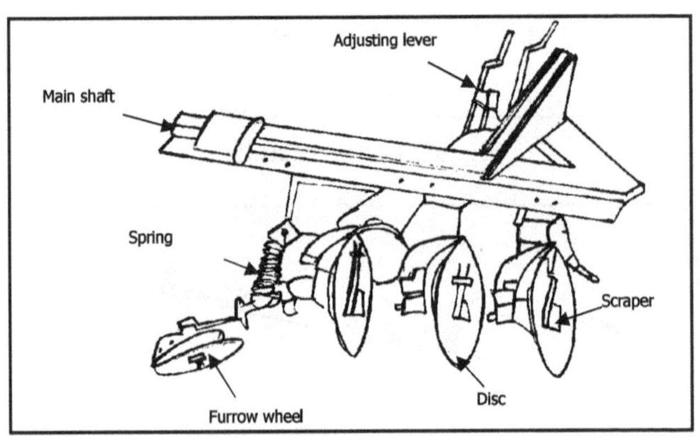

Fig. 11D Disc plough

Fig. 12D Double end bolt

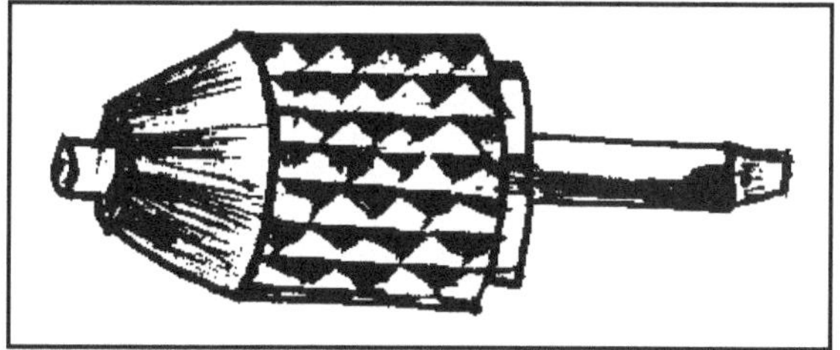

Fig. 13D Drill chuck

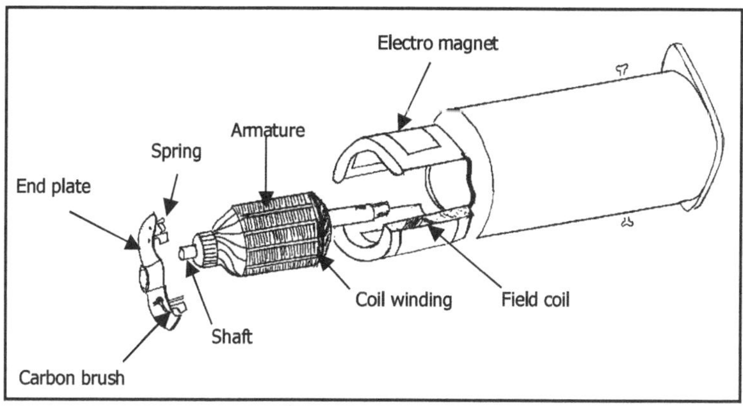

Fig. 14D Dynamo

Ear plug: A device made of a pliable substance, which fits into the ear opening; used to protect an ear from excessive noise or from water.

Ear protector: A device, such as a plug or ear muff, used to protect the human ear from loud noise that may be injurious to hearing, such as that of jet engines.

Earned value: The budgeted cost of the work performed for a given project.

Earth moving: Tillage action and transport operations utilized to loosen, load, carry and unload the soil.

Earthmover: A machine used to excavate, transport, or push earth.

Eccentric: A device used on engines for changing the rotary motion of the crankshaft into a reciprocating motion of the slide valve. This has one circle within another circle but with a different center of rotation. *See* Fig.1E.

Eccentric cam: A cylindrical cam with shaft displaced from the geometric center.

Eccentric gear: A gear whose axis deviates from the geometric center.

Eccentric load: A load imposed on a structural member at some point other than the centroid of the section.

Eccentric rod: The rod, which connects an eccentric with valve mechanism.

Eccentricity: 1. The deviation of the centres of two circles of an eccentric from one another. 2. The radial deviation points on the disc edge in a plane perpendicular to the axis of the disc when disc is rotated about its centre.

Eddy-current clutch: A type of electromagnetic clutch in which torque is transmitted by means of eddy currents induced by a magnetic field set up by a coil carrying direct current in one rotating member.

Eddy-current sensor: A proximity sensor, which uses an altering magnetic field to create eddy current in nearby objects, and then the currents are used to detect the presence of the objects.

Eddy-current tachometer: A type of tachometer in which a rotating permanent magnet induces currents in a spring-mounted metal cylinder; the resulting torque rotates the cylinder and moves its attached pointer in proportion to the speed of the rotating shaft. Also known as drag-type tachometer.

Edge clearance (plough share): The maximum clearance between the cutting edge of share and straight line touching the point of share and wing of share.

Edger: A power grass-cutting machine suitable for cutting lawn and soil, usually in a vertical plane.

Effective field capacity: The actual area covered by an implement based on its total time consumed and its width.

Efficiency: The ratio between an actual result and the theoretically possible result.

Efficiency of a cooking appliance: The ratio of the useful energy transferred to food to the energy consumed by the appliance.

Efficiency of a water heater: The ratio of the energy delivered to the house by hot water to the energy supplied to the water heater.

Efficiency of system: The ratio of the useable work that results from some series of processes to the total amount of energy used during those processes.

Efficiency, thermal: The ratio of the useable work that results from a thermodynamic process (such as a chemical reaction) to the total amount of energy released during the process.

Ejector: A device used to eject the material from the interior of an object.

Elastic: That condition which permits of a deformation when force is applied and a return to original shape on the removal of such force.

Elastic axis: The lengthwise line of a beam along which transverse loads must be applied in order to produce bending only, with no torsion of the beam at any section.

Elastic body: A solid body for which the additional deformation produced by increment of stress completely disappears when the increment is removed. Also known as elastic solid.

Elastic buckling: An abrupt increase in the lateral deflection of a column at a critical load while the stresses acting on the column are purely elastic.

Elastic collision: A collision, in which the sum of the kinetic energies of colliding bodies is the same after and before collision.

Elastic deformation: Reversible alteration of the form or dimensions of a solid body under stress or strain.

Elastic failure: Failure of a body to recover its original size and shape after a stress is removed.

Elastic force: A force arising from the deformation of a solid body which depends only on the body's instantaneous deformation and not on its previous history, and which is conservative.

Elastic hysterisis: Phenomenon exhibited by some solids in which the deformation of the solid depends not only on the stress applied to the solid but also on the previous history of this stress; analogous to magnetic Hysterisis with magnetic induction replaced by stress and strain respectively.

Elastic limit: A limiting point on stress strain curve beyond which the permanent deformation takes place.

Elastic potential energy: Capacity that a body has to do work by virtue of its deformation.

Elastic ratio: The ratio of the elastic limit to the ultimate strength of a solid.

Elastic strength: The greatest stress a bar or a structure is capable of sustaining within the elastic limit.

Elastic vibration: Oscillatory motion of a solid body, which is, sustained by elastic forces and the inertia of the body.

Elasticity: The property whereby a solid material changes its shape and size under action of opposing forces, but recovers its original configuration when the forces are removed.

Elastoplasticity: State of a substance subjected to a stress greater than its elastic limit but not so great as to cause it to rupture, in which it exhibits both elastic and plastic properties.

Electric brake: An actuator in which the actuating force is supplied by current flowing through a solenoid, or through an electromagnet which is thereby attached to disks on the rotating member, actuating the brake shoes; this force is counteracted by the force of a compressing spring. Also known as electromagnetic brake.

Electric car: An automotive vehicle that is propelled by one or more electric motors powered by a special rechargeable electric battery rather than by an internal combustion engine.

Electric drive: A mechanism, which transmits motion from one shaft to another and controls the velocity ratio of the shafts by electrical means.

Electric fence: A fence consisting of one or more lengths of wire energized with high- voltage, low current pulses, and giving a warning shock when touched.

Electric furnace: A furnace, which uses electricity as a source of heat.

Electric hammer: An electric powered hammer, often used for riveting or chalking.

Electric power plant: A power \plant that converts a form of raw energy into electricity, for example, a hydro, steam, diesel, or nuclear generating station for stationary or transportation service.

Electric power system: A complex assemblage of equipment and circuits for generating, transmitting, transforming, and distributing electric energy.

Electric power: The rate of doing work by electricity; measured in watts or kilowatts.

Electric tachometer: An instrument for measuring rotational speed by measuring the output voltage of a generator driven by the rotating unit.

Electric typewriter: A typewriter having an electric motor that provides power for all operations initiated by the touching of the keys.

Electric vehicle: Any ground vehicle whose original source of energy is electric power, such as an electric car or electric locomotive.

Electrical horse power: Equal to 746 watts of electrical energy.

Electrical weighing system; An instrument which weighs an object by measuring the change in resistance caused by the elastic deformation of a mechanical element loaded with the object.

Electricity: A form of energy resulting from a flow of electrons along a conductor.

Electro drill: A drilling machine driven by electric power.

Electro mechanics: The technology of mechanical devices, or processes, which are electrostatically or electro magnetically actuated or controlled.

Electrochemical power generation: The direct conversion of chemical energy to electric energy, as in a battery or fuel cell.

Electrode: 1.A conductor through which electricity enters or leaves an electrolyte. Battery and fuel cell have a negative electrode (the anode) and a positive electrode (the cathode). 2.The part or parts of a resistance-welding machine through which the welding current and pressure are applied directly to the work.

Electrolysis: Decomposition of a material by passing an electric current through it, e.g. the decomposition water into its elemental components (hydrogen and oxygen) through the application of electrical energy.

Electrolyte: The common term applied to the mixture of sulphuric acid and water used in a lead storage cell or battery. All acids, bases, and salts are electrolytes.

Electrolyte level: The point at which the electrolyte should be maintained above the group of plates in the storage battery cell.

Electrolytic copper: Copper produced mainly from the sulphide ores of the Western United States by electrolytic processes. After passing through several processes, it is cast into cakes, slabs, wire bars, ingots, and billets.

Electrolytic grinding: A combined grinding and machining operation in which the abrasive, catholic grinding wheel is in contact with the anodic work piece beneath the surface of an electrolyte.

Electrolytic iron: A very pure iron produced by an electrolytic process. It has excellent magnetic properties and is often used in magnet cores.

Electromagnet: A coil of wire on an iron core actuated by a current of electricity through the coil, which makes a magnet of the ore as long as the current flows through the wire.

Electro-magnetic field: The total magnetic lines of force of the generator or starting motor set up by means of the current flowing through coil windings on soft iron poles.

Electronic fuel injection system (EFI): A system that injects fuel into a spark ignition engine, and uses computer controls to meter the fuel delivery.

Electroplating: The act or processes of depositing metal by electric means.

Electrostatic dusting: The process where electrostatic forces are applied for the deposition of dust.

Element: Any one of the 93 different (simple) substances. Gold, Silver, Lead and Helium are elements.

Elevation: A geometrical projection on a vertical plane, as a front view in mechanical drawing.

Elevator: A conveyor in which the materials are moved upwards.

Elliptical or eccentric gears: Gears, in which the shaft is not in the center. It has almost any shape-oval, heart-shaped, etc. Printing presses usually have good examples of these.

Elongation: The increase in length of a stressed material.

Embossing: The decoration of an object with raised or depressed design.

Embossing hammer: Used for working on the inside surface of hollow objects.

Emergency brake: A hand operated brake commonly known as a parking brake. Rarely used in actual driving except when an automobile is stopped on a road. Its principal use is to prevent motion of an automobile when parked.

Emery: An abrasive compound composed of oxide of alumina, iron, silica, and a small portion of lime. It is used as an

Emery cloth: Powdered emery glued on thin cloth, used for removing file marks and for polishing metallic surfaces.

Emery wheel: A wheel made of emery. It is revolved at a high speed and is used for grinding.

Emission control: Any device or modification added onto or designed into a motor vehicle for the purpose of reducing air-polluting emissions.

Emission standards: Allowable automobile emission levels, set by local, state and federal legislation.

Empirical rule: Any rule or equation which is not deducted from purely mathematical or physical considerations, but which is based upon experience or custom.

Enclosure: A structure whose purpose is to protect equipment from the environment or to provide noise attenuation.

End thrust: The end or longitudinal pressure exerted by a rotating vertical or horizontal shaft. Usually cared for by a thrust bearing.

Endless chain fertilizer distributor: A gravity feed fertilizer distributors fitted with an endless chain running transversely to the direction of travel across the bottom of the hopper. The chain is provided with obliquely mounted fingers, the ends of which push the fertilizer through a slot, which extends across the width of the hopper.

Endless chain mechanism: A combined feed and distributing mechanism consisting of an endless chain with slanting fingers traveling along the bottom of a hopper and pushing fertilizer under a feed gate.

Endothermic reaction: A chemical reaction that draws heat inward during the process of reaction.

Endurance limit: The maximum stress to which material may be subjected without causing failure by fatigue.

Energy: The quantity of work a system or substance is capable of doing, usually measured in British thermal units (BTU) or Joules (J).

Energy availability: Property used to determine the useful work potential of a given amount of energy at some specified state. It is important to realize that energy does not represent the amount of work that a work-producing device will actually deliver upon installation. Rather, it represents the upper limit on the amount of work a device can deliver without violating any thermodynamic laws.

Energy balance: The energy change of a system during a process, which is equal to the difference between the net energy transfer through the system boundary and the energy destroyed within the system boundaries as a result of irreversibility.

Energy content: Amount of energy for a given weight of fuel. Every fuel can liberate a fixed amount of energy when it reacts completely with oxygen to form water. This energy content is measured experimentally and is quantified by a fuel's higher heating value (HHV) and lower heating value (LHV). The difference between the HHV and the LHV is the "heat of vaporization" and represents the amount of energy required to vaporize a liquid fuel into a gaseous fuel, as well as the energy used to convert water to steam.

Energy density: Amount of energy for a given volume of fuel. Thus, energy density is the product of the energy content and the density of a given fuel.

Engine: A machine in which power is applied to do work by the conversion of various forms of energy into mechanical force and motion. It is the prime source of power generation used to propel the machine.

Engine balance: Arrangement and construction of moving parts in reciprocating or rotating machines to reduce dynamic forces, which may result in undesirable vibrations.

Engine cooling: Controlling the temperature of internal combustion engine parts to prevent overheating and to maintain all operating dimensions, clearness, and alignment by a circulating coolant, oil, and a fan.

Engine cycle: Combination of suction, compression, power and exhaust events in an engine. Refer Diesel cycle. Also *Refer* Fig.4D.

Engine cylinder: A cylindrical chamber in an engine in which the energy of the working fluid, in the form of pressure and heat, is converted to mechanical force by performing work on the piston.

Engine displacement: Volume displaced by each piston moving from bottom dead center to top dead center multiplied by the number of cylinders.

Engine driven: Operated by an engine.

Engine efficiency: Ratio between the energy supplied to an engine to the energy output of an engine.

Engine inlet: A place of entrance for engine fuel.

Engine knock: In spark ignition engines, the sound and other effects associated with ignition and rapid combustion of the charge before the flame front reaches it. Also known as combustion knock.

Engine performance: Relationship between power output, revolutions per minute, fuel consumption, and ambient conditions in which an engine operates.

Engine power: Power measured at the crankshaft of the engine with the governor control lever in the position recommended by the manufacturer.

Engine weight per horsepower: The dry weight of an engine divided by the rated horsepower.

Engineer: An expert in design, construction, and development in the fields of mechanism, agriculture, electricity, mining, building, etc.

Engineering: The art and science relating to expert planning and constructing in various fields of industry.

English system: Also known as the *United States Customary System* (USCS), having the unit for mass, length and time as pound-mass (lbm), foot (ft), and second (s) respectively. The pound symbol lb is actually the abbreviation of *libra*, which was the ancient Roman unit of weight.

Enthalpy: It is originated from the Greek word *enthalpien*, which means to heat, is defined as the sum of the internal energy and the product of pressure and change in the volume (P.dv). It is thermodynamic measure of heat content. Enthalpy is used as a way to quantify the amount of energy released or absorbed when a system changes from one state to another.

Enthalpy departure: Is the difference between the enthalpy of a real gas and the enthalpy of the gas at an ideal gas state and it represents the variation of the enthalpy of a gas with pressure at a fixed temperature.

Enthalpy of combustion: The enthalpy of reaction during a steady-flow combustion process when 1 kmol (or 1 kg) of fuel is burned completely at a specified temperature and pressure and represents the amount of heat released.

Enthalpy of formation: The enthalpy of a substance at a specified state due to its chemical composition. The enthalpy of formation of all stable elements (such as O_2, N_2, H_2, and C) has a value of zero at the standard reference

Entropy: A property designated by *S* and is defined as $dS = dQ/T$, where *dS* is change in entropy, dQ is change in heat and T absolute temperature.

Entropy balance relation: The entropy change of a system during a process is equal to the net entropy transfer through the system boundary and the entropy generated within the system as a result of irreversibilities.

Entropy generation: The entropy generated or created during an irreversible process, is due entirely to the presence of irreversibility, and is a measure of the magnitudes of the irreversibility present during that process. Entropy generation is always a positive quantity or zero. Its value depends on the process, and thus it is not a property.

Environment: The region beyond the immediate surroundings whose properties are not affected by the process at any point.

Epicyclic gear: A system of gears in which one or more gears travel around the inside or the outside of another gear whose axis is fixed.

Equation of state: Equation that relates the pressure, temperature, and specific volume of a substance.

Equilibrium moisture content: The moisture content of the product when it is in equilibrium with the surrounding atmosphere.

Equilibrium relative humidity: The relative humidity of the air surrounding the grain, which is in equilibrium with product of given moisture content. The air and product is at the same temp.

Equilibrium: It is the balanced state where there are no unbalanced driving forces within the mechanism.

Equivalence air fuel ratio: The actual air fuel ratio divided by the stoichiometric air fuel ratio. Lean mixtures are described as less than 1.0 and rich mixtures are greater than 1.0.

Equivalent bending moment: A bending moment which, acting along, would produce in a circular shaft a normal stress of the same magnitudes as the maximum normal stress produced by a given bending moment and a given twisting moment acting simultaneously.

Equivalent twisting moment: A twisting moment which, if acting alone, would produce in a circular shaft a shear stress of the same magnitude as the shear stress produced by a given twisting moment and a given bending moment acting simultaneously.

Ergonomics: The scientific study of relationship between man and his working environment. Working environment includes his tools, machines , materials, ambient conditions and physical environment.

Etching: A process of engraving or marking, in which lines are scratched with a needle on a plate or other surface covered with wax, and the parts exposed are subjected to the action of an acid.

Ethanol: An alcohol composed of carbon, hydrogen and oxygen. It is a clear colorless liquid and is the same alcohol found in beer, wine and whiskey. Fermenting a sugar solution with yeast produces ethanol.

Euler equation of motion: A set of the differential equations expressing relations between the force moments, angular velocities, and angular accelerations of a rotating rigid body.

Euler force: The greatest load a long, slender column can carry without buckling, according to the Euler formula for long columns.

Euler formula for long columns: A formula, which gives the greatest axial load that, a long, slender column can carry without buckling, in terms of its length, Young's modulus, and the moment of inertia about an axis along the center of the column.

Evaporation: The process of transferring from a liquid state to a vapour state, such as boiling water to produce steam. Evaporation is the opposite of condensation.

Evaporation tank: A tank used to measure the evaporation of water under controlled conditions.

Evaporative cooling: Lowering the temperature of large mass of liquid by utilizing the latent heat of vaporization of a portion of the liquid.

Evaporator: A device used to vaporize part or all of the solvent from a solution; the valuable product is usually either a solid or concentrated solution of the solute.

Excess air (combustion): Amount of air in a combustion process greater than the amount theoretically required for complete oxidation.

Excess flow valve: A check valve, which permits flow of fluid in either direction but which limits excessive flow in one direction. If the designated flow is exceeded, the valve automatically closes.

Exhaust: 1. The working substance discharged from an engine cylinder or turbine after performing work on the moving parts of the machine. 2. The phase of the engine cycle concerned with this discharge. 3. A duct for the escape of a gases, fumes, and odors from an enclosure, sometimes equipped with an arrangement of fans.

Exhaust emissions: Pollutants emitted into the atmosphere through any opening downstream of the exhaust ports of an engine.

Exhaust gas: Burnt gas leaving an internal combustion engine or gas turbine.

Exhaust-gas analyzer: An instrument that analyzes the gaseous products to determine the effectiveness of the combustion process.

Exhaust gas re-circulation (EGR): A control system that re-circulates a portion of the exhaust gases back into the intake manifold.

Exhaust manifold: The passage from the engine cylinders to the muffler, which conducts the exhaust gases away from the engine. *See* Fig.4A.

Exhaust pipe: The pipe, which extends from exhaust manifold to muffler to carry off exhaust gases. *See* Fig.4A.

Exhaust stroke: The piston stroke (from BDC to TDC) immediately following the power stroke, during which the exhaust valve opens so that the exhaust gases can escape from the cylinder to the exhaust manifold. *Refer* Fig. 9F.

Exhaust system: The system through which exhaust gases leave the vehicle. Consists of the exhaust manifold, exhaust pipe, muffler, tail pipe and resonator.

Exhaust valve: The exit through which the combustion products are expelled from the cylinder. *See* Fig.4A.

Exothermic: A chemical reaction that expels heat outward during the process of reaction.

Expansion: An increase in size. For example, when a metal rod is heated, it increases in length and perhaps also in diameter. Expansion is the opposite of contraction.

Expansion cooling: Cooling of a substance by having it undergoes adiabatic expansion.

Expansion fit: For a right fit of metal parts one within another, the inner part is contracted by dry ice or freezing and placed in position. Opposite of shrink fit.

Expansion joint: a device for overcoming the motion of expansion and contraction in pipes.

Expansion loop: A complete loop installed in pipeline to mitigate the effect of expansion or contraction of the line.

Expansion ratio: The ratio of the volume at which a gas or liquid is stored compared to the volume of the gas or liquid at atmospheric pressure and temperature.

Expansion reamer: A reamer whose diameter may be adjusted between limits by an expanding screw.

Expansion tank (surge tank): A tank that provides room for fuel expansion or heated coolant, giving off any air that may be trapped in the system.

Expansion valve: A valve in which fluid flows under falling pressure and increasing volume.

Experiment: A test or trial for the purpose of determining results under known conditions.

Explosive: Element, which is liable to explode or to violently burst.

Explosive limits: The explosive range of a gas is defined in terms of its lower explosive limit (LEL) and its upper explosive limit (UEL). Between the two limits is the explosive range in which the gas and air are in the right proportions to burn when ignited. Below the LEL there is not enough fuel to burn. Above the UEL there is not enough air to support combustion.

Extension rim: A rim with lugs or cleats fitted to the side of steel driving wheel.

Extension tap: A tap with an extra long shank to permit tapping in places difficult to reach.

Extensive properties: The properties dependent on the size or extent of the system. Mass m, volume V, and total energy E are some examples of extensive properties.

External combustion engine: An engine in which the generation of heat is effected in a furnace or reactor outside the engine cylinder.

External device: A part of an equipment that operates in conjunction with and under the control of a central system, such as a computer or control system, but is not part of the system itself.

External grinding: Grinding the outer surface of a rotating piece of work.

External shoe brake: A friction brake operated by the application of externally contracting elements. *See* Fig.2E.

External thread: The thread on the outside of a screw, or a body.

Extrusion of metal: This is a process by which hot or cold metal is forced under high pressure through an opening to produce a desired shape.

Eyebolt: A bolt provided with a hole or eye at one end, instead of the usual head. The eye receives a pin, stud, or hook, which takes the pull of the bolt. *See* Fig.3E.

Eyeleting: Forming a lip around the rim of a hole.

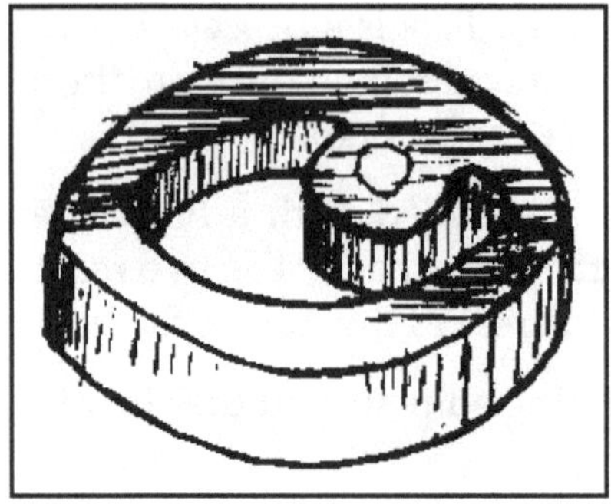

Fig. 1E Eccentric

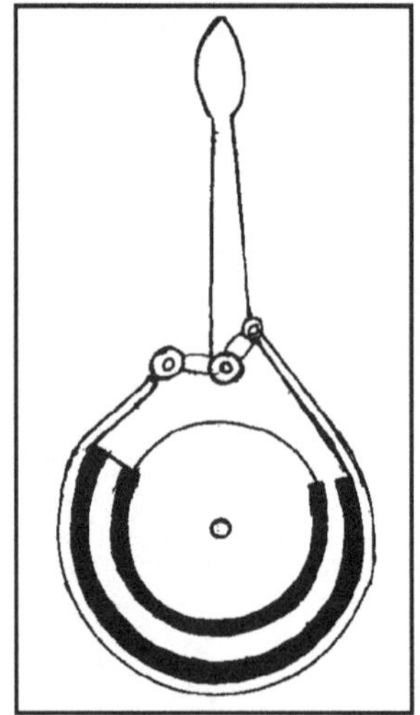

Fig. 2E External shoe brake

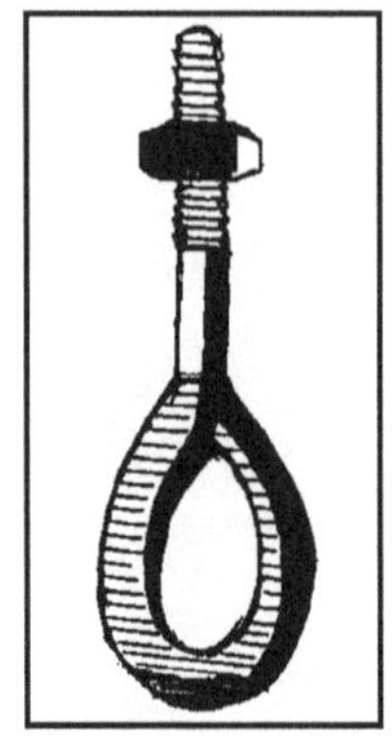

Fig. 3E Eyebolt

Fabrication: The act of building or putting together. Forming into a whole by uniting or assembling of parts.

Face lathe: A short-bed, deep gap lathe designed for the machining of large, flat surfaces.

Face milling: Milling flat surfaces perpendicular to the rotational axis of the cutting tool.

Faceplate or face chuck: A circular plate for attachment to the spindle in the headstock of a lathe. Work may be clamped or bolted to it. The slots engage the tail of the lathe dog.

Factor of safety: The number, which results by dividing the ultimate strength of a material by the actual stress on a sectional area.

Factor of stress intensity: The ratio of the maximum stress to which a structural member can be subjected to the maximum stress of safety.

Fahrenheit: Temperature scale and unit of temperature (°F). Named for German physicist Gabriel Daniel Fahrenheit (1686 -1736) who was the first to use mercury as a thermometric fluid in 1714. On the Fahrenheit scale, the ice and steam points are assigned 32 and 212 °F.

Fahrenheit's hydrometer: A type of hydrometer which carries a pan at its upper end in which weights are placed; the relative density of a liquid is measured by determining the weights necessary to sink the instrument to a fixed mark, first in water and then in the liquid being studied.

Fail-safe control: A control so designed that a malfunction of any of its components shall automatically stop the operation of the device or equipment controlled by it.

Failure: The inability of materials and structures to endure or accomplish the work for which they were selected and designed.

False key: A round pin driven into a hole drilled one half in the end of a spindle and the other half into a boss.

False work: Usually temporary framework, bracing, or support used as an aid in construction and removed when construction is completed.

Fan efficiency: The ratio obtained by dividing a fan's useful power output by the power input (the power supplied to the fan shaft); it is expressed as percentage.

Fan nozzle: A hydraulic spray nozzle, which produces a narrow elliptical spray pattern.

Fan pulley: A pulley attached to the fan to engage the belt, which drives it, and on many cars also the circulating pump and the generator.

Fan rating: The head, quality, power, and efficiency expected from a fan operating at peak efficiency.

Fan ring: Circular metallic collar encircling (but spaced away from) the tips of the fan blade in process equipment, such as air-cooler heat exchangers. Ring design is critical to the efficiency of fan performance.

Fan: 1.An appliance for producing airflow by means of blade fixed to a rotating shaft. 2. A device that increases the pressure of a gas slightly and is mainly used to mobilize a gas.

Farmyard manure spreader: A machine to carry farmyard manure and spread it in a regulated quantity.

Fast coupling: A flexible geared coupling that uses two interior hubs on the shafts with circumferential gear teeth surrounded by a casing having internal gear teeth to mesh and connect the two hubs.

Fast pulley: A pulley employed to transmit motion, being attached to its shaft by a setscrew or a key.

Fastener: A device for joining two separate parts of an article or structure.

Fatigue factor: The element of physical and mental exhaustion in a time- motion study; the multiplier used to add the fatigue allowance to the normal time.

Fatigue life (material) : The number of applied repeated stress cycles a material can endure before failure.

Fatigue limit (material): The maximum stress the material can endure for an infinite number of stress cycles without braking.

Fatigue of material: Material, which has been long subjected to severe or moderate straining, deteriorates in strength, and will brake under loads previously sustained with safety.

Fatigue ratio: The ratio of the fatigue limit or fatigue strength to the static tensile strength.

Fatigue strength: The maximum stress a material can endure for a given number of stress cycles without braking. Also known as endurance strength.

Fatigue test: Test to determine the range of altering stress, which a material can withstand without braking.

Feather - edge: A keen edge, tapering off to nothing.

Feather joint: A joint made by cutting a mating groove in each of the pieces to be joined and inserting a feather in the opening formed when the pieces are butted together. Also known as ploughed and tongued joint. One of a series of joints in a fault zone formed by shears and tension. Also known as pinnate joint.

Feather or sunk key: A parallel key, which is partly sunk into a recess in its shaft, so as to form an integral part of the shaft. The keyway in the wheel or clutch carried on the shaft is made large enough to permit these parts to slide longitudinally on the shaft. *See* Fig.1F.

Feed control valve: A small valve, usually a needle valve, on the outlet of the hydraulic feed cylinder on the swivel head of a diamond drill, used to control minutely the seed of the hydraulic piston travel and hence the rate at which the bit is made to penetrate the rock.

Feed gate: An adjustable shutter to control the flow of material.

Feed gears: Gears used to drive the feed rod and control the rate of feed.

Feed mechanism (fertilizer) **:** A mechanism for discharging the fertilizer from hopper at a predetermined rate. Different types of fertilizer feed mechanisms as serrated disc type, star wheel type and spur wheel type etc are available. *See* Fig.2F.

Feed mechanism (lathe machine): Consists of an arrangement of gears, screws or other devices for controlling the feed of the tool to the work or the work to the tool.

Feeder beater (thresher): In front of threshing cylinder and rotating in the same direction, the beater is provided to aid in the stripping the crop from the conveyor and feeding it to the cylinder.

Feeder conveyor: The auxiliary conveyor to assist in feeding the crop to the threshing cylinder.

Feeding (thresher) **:** The operation of the conveying cut crop into threshing unit.

Feeler: Gauge for determining the size of a piece of work, the accuracy of the test depending on the sense of touch.

Feller: Self- propeller machine design to fell standing trees.

Feller buncher: Self- propelled machine designed to fell standing trees and arrange them in bunches on the ground.

Feller forward: Self- propelled, self-loading machine designed to fell the standing trees and move the felled trees by carrying them.

Feller skidder: Self- propelled machine designed to fell standing trees and transport them by dragging.

Felling: Cutting the standing stem above the root system.

Felling sights: Marks on the chain saw to aid felling the trees in a desired direction.

Female portion of work: The recessed portion of any piece of work into which another part fits is called the female portion.

Female thread: A thread, which is cut in a hole or on a hollow surface.

Ferromanganese: An alloy of 20 to 80 % of manganese and 5 - 7 % of carbon in powder form is added to the molten metal in the ladle. Acts as a deoxidizing agent and counteracts the influence of an excess of sulphur. Used specially when the mix contains a large proportion of steel scrap.

Ferro nickel: A nickel steel alloy used for rheostat and coils.

Ferro phosphorus: An iron of high phosphorus contain used in making steel for tinplate. Also called as 'Fosforus'.

Ferro silicon (metal) : A hard steel containing 97.6% iron, 2% silicon and 0.4% carbon.

Ferrule: A metal ring or cap attached to the end of the tool handle, post or other device to strengthen and protect it.

Fertilizer attachment: A detachable device, which is fitted to a seed drill to apply fertilizer at the time of sowing. *See* Fig.15F.

Fertilizer broadcaster: A fertilizer distributor with a spreading width substantially greater than the width of machine.

Fertilizer distributor: A machine, which distributes fertilizer at regulated and selected rate.

Fertilizer drill: A Machine to deposit fertilizer in soil at regulated and selected rates and at pre-determined depth.

Fertilizer gun: A Machine to throw a jet of fertilizer by mechanical, pneumatic or other means, usually fitted to the side of the machine

Fertilizing: Process of fertilizer application in or on the soil, or on the plants.

F-head Engine: A combination of L-head and I head types. Intake valves are overhead and the exhaust valves are in cylinder blocks. This style engine is not much used.

Field (magnetic)**:** Usually refers to the magnetic lines of force or flux flowing between pole shoes of the electro-magnet.

Field capacity: 1. Area of the field covered by a tool, implement or machine per unit time. It is expressed in hectare/hour, acre/day etc. 2.The amount of water retained in a soil against a force of gravity at any specified time (about 24 to 36 hours) after flooding.

Field ditch: A smaller ditch constructed within a field either for irrigation or for drainage.

Field efficiency: The quotient of the effective field capacity and theoretical field capacity expressed in percentage.

Field heap spreader: A manure spreader for spreading heaps arranged in the rows.

Field system: The part of a farm system, which covers one field or area for which it is designed.

Filament(electrical)**:** The small electrical wires in the electric bulb, which become heated to incandescent (glowing)

point when electric current is allowed to flow through them; usually, made of tungsten.

File hard metal: When a metal is so hard that it cannot be filed it is said to be 'file hard'.

File hardness: Hardness of a material as determined by testing with a file of standardized hardness; a material, which cannot be cut with a file, is considered as hard as or harder than the file.

Filing: The removal of material, finishing\g, and/ or fitting by use of a file.

Fill: In engineering, the material used to bring a low spot up to designed grade.

Filled system thermometer: A thermometer, which has bourdon tube, connected by a capillary tube to a hollow bulb; the deformation of the bourdon tube depends on the pressure of a gas (usually nitrogen or helium) or on the volume of a liquid filling the system. Also known as filled thermometer.

Filler plug: The plug in any housing in the automobile, which can be removed for checking the grease or lubricant level within the housing. The opening thus provided is used for filling the housing with lubricant to the level of the filler plug opening.

Fillet: A concave curve connecting two surfaces, which meet at an angle. By avoiding sharp angles, it adds to the strength and beauty of a design.

Fillet gage: A gage for measuring convex or concave surfaces.

Fillister: A cylindrical head of a cap screw slotted for a screwdriver.

Filter: 1. A device to remove dirt from oil, air or water. 2.A device through which air, gases, or liquids are passed to remove impurities.

Filter element: The active part of an oil filter consisting of cloth or other material. Its purpose is to remove grit, dirt, and other foreign matter from the motor oil.

Filter press: A metal frame on which iron plates are suspended and pressed together by a screw device; liquid to be filtered is pumped into canvas bags between the plates, and the screw is tightened so that pressure is furnished for filtration.

Filter pump: An aspirator or vacuum pump, which creates a negative pressure on the filtrate side of the filter to hasten the process of filtering.

Filter screen: A fine pored medium through which a liquid will pass and on which solids deposit; the medium may be a metal sieve screen or a woven fabric of metal or of natural or synthetic fibers.

Filterability: The adaptability of a liquid-solid system to filtration; system is not filterable if it is too viscous to be forced through a filter medium, or if the solids are too small to be stopped by the filter medium.

Final aspiration: the suction or air blast which removes light seeds of poor germination or trash from the screened bold seed mass after it has passed over the last screen of the seed cleaner.

Final drive: A device provided for additional reduction of speed between the drive shaft from the transmission and the axle connecting the drive member. *See* Fig.3F.

Fineness modulus: A number denoting the fineness of a fine aggregate or other fine material such as sand or paint.

Finger(cutter bar) **:** A finger like attachment bolted to the forward edge of a guard bar and through which knife section reciprocates. *Refer* Fig. 21C.

Finger bar (cutter bar)**:** A steel bar to carry guards wearing plates and knife guide. *Refer* Fig. 21C.

Finish grinding: The last action of a grinding operation to achieve a good finish and accurate dimensions.

Fire crack: A crack resulting from thermal stress, which propagates on the heated side of a shell or header in a boiler or a heat transfer surface.

Fire detector: A temperature-sensing device designed to sound an alarm, to turn on a sprinkler system, or to activate some other fire preventive measure at the first signs of fire.

Fire door: 1. The door or opening through which fuel is supplied to a furnace or stove, 2. A door that can be closed to prevent the spreading of fire, as through a building or mine.

Fire hose: A collapsible, flameproof hose that can be attached to a hydrant, standpipe, or similar outlet to supply water to extinguish a fire.

Fire line: A pipe work system dedicated to provide water for extinguishing fire.

Fire room: That portion of a fossil fuel burning plant, which contains the furnace and associated equipment.

Firebox: The furnace of a locomotive or similar type of fire tube boiler.

Firecracker: A cylindrically shaped item containing an explosive and a fuse; used to simulate the noise of an explosive charge.

Firing order (FO): The sequence of order in which cylinders in the engine are fired. Or the sequence in which the power stroke in each cylinder of an engine occurs. For four cylinder engines the most commonly used firing orders are 1-3-4-2 and 1-2-4-3. And for six cylinder engine firing order may be 1-4-2-6-3-5 or 1-5-3-6-2-4.

Firing interval (FI): The interval between successive power strokes in different cylinders of an engine.

$$F.I. = 720^0 / \text{no. of cylinders} - \text{for four stroke engine, and}$$

$$F.I. = 360^0 / \text{no. of cylinders} - \text{for two stroke engine}$$

Firing pressure: The highest pressure in an engine cylinder during combustion.

Firmer chisel: A small hand chisels with a flat blade; used in woodworking.

First cost: The sum of the initial expenditures involved in capitalizing a property; includes items such as transportation, installation, preparation for service, as well as other related costs.

First law of thermodynamics: Whenever a system undergoes a cyclic change, the algebraic sum of work transfer is proportional to the algebraic sum of heat transfer. Or heat and work are mutually convertible one into other.

First gear: The gear connection, which is used for lowest speed, often spoken of as low gear.

First-class lever: A lever with fulcrum between load and effort.

Fishing tool: A device for retrieving objects from inaccessible locations.

Fishplate: A plate of metal covering the butt joints of boilers, rails, and other work.

Fishtail cutter: A tool for cutting grooves or seats in shafts; suitable for light cut and feed.

Fitting: The bringing together of engines, machines, etc., after they have left the hands of the machine workers.

Fixed cost: A cost that remains unchanged during short - term changes in production level. Also, known as overhead; overhead cost.

Fixed end: An end of a structure, such as a beam, that is clamped in place so that both its position and orientation are fixed.

Fixed linkage system: Linkage formed between the skeletal elements of a human and fixed machine in a human - machine system.

Fixed platform trailer: An agricultural trailer in which the platform is fixed with the chassis

Fixing moment: The bending moment at the end support of a beam necessary to fix it and prevent rotation. Also known as fixed end moment.

Flail mower: A grass-cutting machine, which utilizes a power source to rotate a horizontal shaft with blade. Cutting action is accomplished by impact with rotating blades.

Flame detector: A sensing device, which indicates whether or not a fuel is burning, or if ignition has been lost, by transmitting a signal to control system.

Flame gun: An appliance, which produces a burning flame under pressure to control, weeds, pests etc.

Flame speed: The sum of burning speed and displacement velocity of the unburnt gas mixture.

Flame temperature: The temperature of a flame burning a stoichiometric mixture of fuel and air (neither fuel nor air is in excess).

Flamethrower: A device used to project ignited fuel from a nozzle so as to cause casualties to personnel or to destroy material such as weeds or insects.

Flammability limits: The flammability range of a gas is defined in terms of its lower flammability limit (LFL) and its upper flammability limit (UFL). Between the two limits is the flammable range in which the gas and air are in the right proportions to burn when ignited. Below the lower flammability limit there is not enough fuel to burn. Above the higher flammability limit there is not enough air to support combustion.

Flange: A rib or rim added for strength. Also used for guiding or attaching another object (component).

Flanged pipe: A pipe with flanges at the ends; can be bolted end to end to another pipe.

Flank: The side of gear teeth lying below the pitch line. *Refer* Fig.9G.

Flap trap: In plumbing a trap fitted with a hinged flap that permits flow in one direction only, thus preventing backflow.

Flap valve: A valve fitted with a hinged flap or disk that swings in one direction.

Flash boiler: A boiler with hot tubes of small capacity; designed to immediately convert small amounts of water to superheated steam.

Flash line: A raised line on the surface of a molding where the mold faces joined.

Flash mold: A mold, which permits excess material to escape during closing.

Flash point: The temperature at which the fuel produces enough vapors to form an ignitable mixture with air at its surface.

Flashing ring: A ring around pipe that holds it in place as it passes through a partition such as a floor or wall.

Flat belt: A power transmission belt, in the form of leather belting, used where high-speed motion rather than power is the main concern.

Flat belt pulley: A smooth, flat-faced pulley made of cast iron, steel, wood, paper, etc and is used with a flat-belt drive.

Flat chisel: A chisel used for obtaining a flat surface on metal by chipping.

Flat drill: A type of drill, its cutting blade having two parallel, beveled edges. Used for drilling out cored holes.

Flat roller: Roller consisting of one or more hollow or solid cylinders with a smooth surface. *See* Fig.4F.

Flat rope: A steel or fiber rope having a flat cross section and composed of a number of loosely twisted ropes placed side by side, the lay of the adjacent stands being in opposite directions to secure uniformity in wear and to prevent twisting during winding.

Flat spray: Spray with a flat shape.

Flat trajectory: A trajectory, which is relatively flat, that is, described by a projectile of relatively high velocity.

Flatbed plotter: A graphics output device that draws by moving a pen in both horizontal and vertical direction over a sheet of paper; the overall size of the drawing is limited by the height and width of this bed.

Flat-blade turbine: An impeller with flat blades.

Flat-flamed burner: A burner, which emits a mixture of fuel and air in a flat steam through a rectangular nozzle.

Flathead rivet: A small rivet with a flat manufactured head used for general-purpose riveting.

Flatter: A kind of hammer used by blacksmith.

Flat-tube radiator: Made from flat tubes which, instead of running direct from tank to tank, are bend in such manner that they will be two or three times as long as a straight tube used for the same purpose. This bending retards the flow of water and presents greater cooling surface. It is frequently spoken of as a Honeycomb Radiator, but is not the true honeycomb type.

Flaw: A crack or fracture in casting or forging. In general, any defect which may eventually cause failure.

Fleam: The angle of bevel of the edge of the teeth of a saw with respect to the plane of the blade.

Fleet: 1.Sidewise movement of a rope or cable when winding on a drum. 2. An organization of ships, aircraft, marine forces, and shore-based fleet activities, all under a commander who may exercise operational as well as administrative control. 3. All naval operating forces.

Flexibility: The quality of state of being able to be fixed or bent repeatedly.

Flexible coupling: A flexible ball coupling consists of two disks attached to the shaft ends, hollowed on their faces to encircle a ball placed between them.

Flexible disc transplanter: A transplanter, the planting element of which consists of flexible discs.

Flexible shaft: A shaft made of jointed links encased in flexible tubing; used to transmit power in places where a straight shaft could not be used.

Flexible tine harrow: A harrow consisting of network of spring tines. *See* Fig.5F.

Float (machinery) **:** An implement used for packing and smoothening the surface of the soil. *See* Fig.6F.

Float (sleigh)**:** A component in front of the weeding roll, which ensure an easy sliding motion during operation through better floatation of the weeder.

Float gage: Any one of several types of instruments in which the level of a liquid is determined from the height of a body floating on its surface, by using pullies, levers, or other mechanical devices.

Float trap (boiler) **:** A valve actuated by a hollow metal float so arranged that condensation and air may pass but a steam will be held.

Float type rain gage: A class of rain gage in which the level of the collected rainwater is measured by the position of a float resting on the surface of the water; frequently used as a recording rain gage by connecting the float through a linkage to a pen which records on a clock driven chart.

Float valve: A valve whose on-off action is controlled directly by the fall or rise of a float concurrent with the fall or rise of liquid level in a liquid containing vessel.

Floating axle: An axle on which the shaft is relived of all loads or stresses except turning the wheel.

Floating pan: An evaporation pan in which the evaporation is measured from water in a pan floating in a larger body of water.

Floating piston pin: A piston pin, which is not locked in the connecting rod or the piston, but is free to turn or oscillate in both the connecting rod and the piston.

Floating tool: A tool so secured in its holder that it might be guided in its operation by the piece on which it works.

Flooding (carburetor)**:** An excessive amount of rich fuel mixture being fed to the engine resulting in difficult starting.

Flow chart: A graphical representation of the progress of a system for the definition, analysis, or solution of a data processing or manufacturing problem in which symbols are used to represent operations, data or material flow, interrelationships among the components etc.

Flow coat: A coating formed by pouring a liquid material over the object and allowing it to flow over the surface and drain off.

Flow control: Any system used to control the flow of gases, vapors, liquids, slurries, pastes, or solids particles through or along conduits or channels.

Flow control valve: A valve whose flow opening is controlled by the rate of flow of the fluid through it; usually controlled by differential pressure across an orifice at the valve. Also known as rate-of-flow control valve.

Flow direction: The antecedent- to- successor relation, indicated by arrows or other conventions, between operations on a flow chart.

Flow line: 1. The connecting line or arrow between symbols on a flow chart or block diagram. 2. Mark on a molded plastic or metal article made by the meeting of two input flow fronts during molding. Also known as weld line.

Flow measurement: The determination of the quantity of a fluid, a liquid, a vapour, or a gas, that passes through a pipe, duct, or open channel.

Flow meter: An instrument used to measure pressure, flow rate, and discharge rate of a liquid, vapour, or gas flowing in a pipe. Also known as fluid meter.

Flow process: System in which fluids are handled in continuous movement during chemical or physical processing or manufacturing.

Flow soldering: Soldering of printed circuit by moving them over a flowing wave of molten solder in a solder bath; the process permits precise control of the depth of immersion in the molten solder and minimizes heating of the board. Also known as wave soldering.

Flow valve: A valve that closes itself when the flow of a fluid exceeds a particular value.

Flow work: Work required to push mass into or out of control volumes. On a unit mass basis this work is equivalent to the product of the pressure (P) and specific volume of the mass(v), Pv.

Flowchart symbol: Any of the existing symbols normally used to represent operations, data or materials flow, or equipment in a data-processing problem or manufacturing process description.

Flue gas: Gaseous combustion products from a furnace.

Flue gas expander: In a petroleum processing system, a turbine for recovering energy at the point where combustion gases are discharged under pressure to the atmosphere; the reduction in pressure drives the turbine impeller.

Fluid coupling: A type of clutch arrangement between two working parts; one of which drives the other by means of fluid. *See* Fig.7F.

Fluid drive: The fluid coupling used in place of, or as an aid to, a friction clutch. A hydraulic means of power transfer from the driving to the driven components of the automobile.

Fluid friction: When the particles of a fluid in motion and the outer surfaces of the fluid are in contact with solid surfaces the fluid body is divided into numerous layers within itself. The friction produced by the slipping of these layers over one another and by rubbing effect between the molecules of the fluid is called fluid friction.

Fluid mechanics: The science concerned with fluids either at rest or in motion, and dealing with pressures, velocities and accelerations in the fluid, including fluid deformation and compression or expansion.

Fluid ton: A unit of volume equal to 32 cubic feet or approximately 9.061×10^2 cubic meters; used for many hydrometallurgical, hydraulic, and other industrial purposes.

Fluid transmission: Automotive transmission with fluid drives.

Flume: An open channel constructed of steel, reinforced concrete, or wood and used to convey water.

Flush: Parts are said to be flush when their surface are on the same level.

Flush bolt: A bolt whose head is let into a counter bored hole so that the top of its head rests level with the face of the plate into which it is sunk.

Flute: The concave channel in a reamer, tap, or drill.

Fluted feed roller: A seed-metering device with adjustable fluted roller to collect and deliver seeds into a seed tube. *See* Fig.8F.

Fluting: A machining operation whereby flutes are formed parallel to the main axis of cylindrical or conical parts.

Flutter or bounce: In engine valves, refers to a condition where the valve is not held tightly on its seat during the time the cam is not lifting it.

Flutter valve: A valve that is operated by fluctuations in pressure of the material flowing over it; used in carburetors.

Flux: Magnetic field (lines of force) about the end of an electromagnet of permanent magnet; in the generator the lines of force flowing between pole shoes.

Flywheel: An energy storage device (heavy wheel) in which a balanced mass spinning around a constant axis stores energy by means of momentum as rotational kinetic energy.

Flywheel magnet: A magnet having a high tension magnet built into the flywheel of the engine with other necessary ignition parts built into the hollowed cut portion of the flywheel, although usually not attach thereto.

Flywheel marking: Marks on the face of a flywheel, which serve as a guide for the proper timing of valve action in an engine.

Flywheel type chaff cutter: A chaff cutter having rotating flywheel with blades.

Focusing (head lamp)**:** The art of adjusting the position of the light bulb in a head lamp with reference to the center of the parabolic reflection so that the lamp will throw the correct beam of light

Folding rule: A collapsible instrument used for measuring.

Follower: A wheel, which is driven by another wheel. *Refer* Fig. 3C.

Foot brake: Brake operated by foot action as distinguished from the hand or parking brake.

Foot lever: A lever worked by the pressure of the foot alone.

Foot valve: 1. A valve in the bottom of the suction pipe of a pump, which prevents backward flow of fluid. 2. Check valve opening upwards installed at the inlet end of the suction pipe to retain water in the pump for priming.

Foot-pound: 1. Unit of energy or work in the English system, equal to the work done by 1 pound of force when the point at which the force is applied is displaced 1 foot in the direction of the force; equal to approximately 1.3558 joule. abbreviated as ft-lb or ft-lbf. 2. Unit of torque in the English system, equal to the torque produced by 1 pound of force acting at a perpendicular distance of 1 foot from an axis of rotation. Also known as pound-foot, abbreviated as lb-ft.

Footstep or footstep bearing: A bearing used at the lower end of a vertical shaft or spindle to carry the end thrust.

Forage harvester: A machine to cut or pick up, chop or lacerate fodder and deliver it to a vehicle.

Force: That which changes, or tends to change the state of rest or motion of the body acted upon. It is measured in Newton or kgf. Force has three characteristics; direction, place of application and magnitude.

Force constant: The ratio of the force to the deformation of a system whose deformation is proportional to the applied force.

Force polygon: A closed polygon whose sides are vectors representing the forces acting on a body in equilibrium.

Force pump: A pump in which the water is lifted by the force due to atmospheric pressure acting against a vacuum.

Force-feed drill: A seed drill in which a positive metering mechanism comprises fluted rollers or internal double run wheels. *See* Fig. 8F and *Refer* Fig. 3I.

Forced feed lubrication system: In this system, oil is pumped directly to the crankshaft, connecting rod, piston pin, timing gears, crankshaft of engine etc of an engine. Oil pump used is gear type or vane type positive displacement pump. This system is used on high-speed multicylinder engines as tractor, truck etc.

Foreign body: Foreign objects including pieces of tiles, pipes, wires, or crushed rocks that occurs in or have been inserted in soil.

Foreign element: A work element, which is not a part of the normal work cycle, either because it is accidental or because it occurs only occasionally.

Foreign materials (soil): All foreign matter in soil including residues, soil additives and foreign bodies that have not originated in the development of soil.

Foreman: A man in charge of a group of workmen. He is usually responsible to a superintendent or manager.

Forester's shear: A shear for pruning thicker stock.

Forging press: A machine used to exert the pressure needed in the forging.

Forgings: Pieces of masses of metal shaped by hammering.

Form grinding: Grinding by use of a wheel whose cutting face is contoured to the reverse shape of the desired form.

Forming die: A die like a drawing die, but without a blank holder.

Forming tools: Tools with their working or cutting edges shaped like the form to be produced on the work.

Forwarder (tree falling): Self-propeller machine, usually self-loading designed to move trees or the parts of trees by carrying them completely off the ground. Normally forwarders are used for off-road transportation.

Forwarding: Moving trees or parts of the trees out of the forest by carrying them.

Foundation bolts: Bolts used for holding down or anchoring machinery or structural parts to the foundation on which they rest.

Foundry: A building or place in which metal castings are made.

Four bar linkages: A plane linkage consisting of four links pinned tail to head in a closed loop with lower or closed joints.

Four wheel drive: A drive having live axles front and rear, driving power being delivered to all four wheels.

Four-cycle engine: Also known as Otto cycle, where an explosion occurs every at every half revolution of the crankshaft. *Refer* Otto cycle.

Four-stroke cycle: A power cycle concluded in four strokes of the engine. In the auto engine, the first down stroke is the intake/suction; the next upstroke allows compression; ignition and expansion/power on the third stroke, and exhaust on the fourth. *See* Fig.9F.

Four-stroke engine: An internal combustion engine, in which the piston executes four complete strokes (two mechanical cycles) within the cylinder, and the crankshaft completes two revolutions for each thermodynamic cycle.

Four-way valve: A valve at the junction of four waterways, which allows passage between any two adjacent waterways by means of a movable element operated by a quarter turn.

Fraction defective: The number of units per 100 pieces, which are defective in a lot; expressed as a decimal.

Fracture: To brake apart; to separate the continuous parts of an object by sudden shock or by excessive strain.

Fragments: Masses of soil, which are created by fracture along natural surfaces of weakness.

Framed plough: A plough the body (or bodies) of which is (are) attached to a framed beam.

Framework: The load-carrying frame of a structure may be of timber, steel or concrete.

Free drop: To airdrop supplies or equipment without parachute.

Free end: In a cantilever, the end, which is not fixed, is always called the 'free end'.

Free hand: Executed with the hand without the aid of drawing instruments.

Free joint: A robotic articulation that has six degrees of freedom.

Free light: Unconstrained or unassisted light.

Free light trajectory: The path of a body in free fall.

Freeze: 1. To solidify a liquid by removal of heat. 2.To permit drilling tools, casing, drive pipe, or dill roads to become lodged in a borehole by reason of caving walls or impaction of sand, mud, or drill cuttings, to the extent that they cannot be pulled out. Also known as bind-seize.

Freezer: An insulated unit, compartment, or room in which perishable foods are quick-frozen and stored.

Freezing point: The temperature at which crystals of hydrocarbons formed on cooling disappear when the temperature of the fuel is allowed to rise.

Freight car: A railroad car on which freight is transported.

Freighter: A ship or aircraft used mainly for carrying freight.

French: A unit of length used to measure small diameters, especially those of fiber optic bundles, equal to 1/3 millimeter.

Friction: Holding force or resistance to motion between two surfaces in contact with each other.

Friction brake: A brake kin, which the resistance is provided by friction.

Friction clutch: A coupling for connecting two working parts, one of which is driven by source of power and the other of which is driven by the clutch through the friction of the plate and friction facings against each other. Friction clutches are divided in three subdivisions as 1. Single plate clutch, 2. Multiple plate clutch and 3. Cone clutch. Refer single plate clutch, multiple plate clutch and cone clutch.

Friction coupling: Any one of a variety of couplings, which operate through frictional contact.

Friction disk: The disk of a friction drive.

Friction gear: Gear in which motion is transmitted through friction between two surfaces in rolling contact.

Friction loss: Mechanical energy loosed because of mechanical friction between moving parts of a machine.

Friction of motion: That friction which must be overcome by force in order to keep any solid body (spherical or cylindrical) moving over a plane surface after it is one's set in motion.

Friction of rest: That friction which must be overcome by a force in order to start any solid body (spherical or cylindrical) sliding or rolling over a plane surface.

Friction polisher (grain) : The type of polisher, which operates on the principle of removing the bran from the kernel by using friction between the rice kernels and the screen in the polisher.

Friction saw: A toothless circular saw used to cut materials by fusion due to frictional heat.

Frictional horse power (FHP): It is the power required to run the engine at a given speed without producing any useful work. It represents the friction and pumping losses of an engine.

Friction wheel: Any wheel, which drives or is driven by friction, as when contact takes place only between smooth or grooved surfaces.

Frog (plough) : The part to which other components of the plough bottom are attached.

Front axle tramp: The condition existing when the ends of the front axle move up and down alternately causes first one wheel to tend to leave the road and then the other. This condition is a danger to safe performance of the vehicle.

Front wheel drive: A construction in hitch in which the live propelling axle is at the front end. The rear axle is dead.

Front-end geometry: The mathematical science, which deals with the angles, lines and points of front system of axle design with reference to safe and satisfactory steering.

Fuel: 1.A substance that releases energy when reacted chemically with oxygen.2.combustion matter, such as wood, coal, gas, or oil, which may be used to feed a fire or operate an engine.

Fuel-air ratio: The reciprocal of air-fuel ratio.

Fuel cell: The chemical energy of the fuel is directly converted to electric energy, and electrons are exchanged through conductor wires connected to a load. It convert chemical energy to electric energy essentially in an isothermal manner.

Fuel filter: A device, as an internal combustion engine, that removes dirt and solid particles from the fuel oil. The filtering element consists of metal wires in conjunction with various media such as packed fibers, woven cloth, felt, paper etc. These filters are replaced at certain intervals specified by the manufacturer. *See* Fig.10F.

Fuel gauge: A gauge that indicates the amount of fuel in the fuel tank or cylinder.

Fuel injection: The delivery of fuel to an internal combustion engine cylinder by pressure from mechanical pump.

Fuel injection system: A system (replacing the conventional carburetor) that delivers fuel under pressure into the combustion chamber, pre-combustion chamber, turbulence chamber, or into the airflow just as it enters each individual cylinder. *See* Fig.11F.

Fuel injector: A pump mechanism that sprays fuel into the cylinder of an internal combustion engine at the appropriate part of the cycle. *See* Fig.11F, 13F.

Fuel knock: *Refer* detonation.

Fuel level indicator: An instrument for indicating the amount of fuel in a tank.

Fuel pump: A device operated by a piston or diaphragm to create vacuum, which insures supply of fuel to carburetor or mixing chamber.

Fuel system: The system (fuel cylinders and lines, gauge, fuel pump, carburetor, and intake manifold) that delivers the combustible mixture of vaporized fuel and air to the engine cylinders. *See* Fig.11F, 12F.

Fuel tank: The storage tank for fuel on a vehicle. *See* Fig.11F, 12F.

Fulcrum: The point on which a lever turns.

Fulgurator: An atomizer used to spray salt solutions into a flame for analysis.

Full trailer: A towed vehicle whose weight rests completely on its own wheels.

Full width distributor (fertilizer): A fertilizer distributor in which the fertilizer falls to the ground across the entire width of the spreader.

Fuller: To form grooves, as in blacksmith's work.

Fumigator: An appliance to generate and distribute gases or smokes.

Funnel: 1. A hollow cone shaped vessel with spout at the end to permit free pouring from one container to another.2. A vent for ventilation as used on steamships.

Furlong: A measure of length equal to 1/8 of a mile, 660 ft.

Furnace: An apparatus in which heat is librated and transferred directly or indirectly to a solid or fluid mass for the purpose of effecting a physical or chemical change.

Furniture: Useful and decorative movable articles, such as tables, chairs, etc. placed in a building. Tool racks, lathe pans, tote boxes etc.

Furrow: The trench formed by a tool in the soil during operation.

Furrow crown: The peak of turned furrow slice.

Furrow irrigation: This method consists in making the land into ridges and furrows, and irrigating the area through furrows.

Furrow opener: A part of seed drill for opening a furrow and assisting in placing of seeds *See* Fig.14F.

Furrow planting: Planting in the bottom of furrows.

Furrow slice: The soil mass cut and turned by tool

Furrow sole: The bottom surface of the furrow.

Furrow wall: The undisturbed wall of the furrow

Fuse: A link of soft metal housed in a glass tube metal capped at each end; its purpose is to protect an electric circuit against too great a flow of current by melting out when the current flow reaches a certain point.

Fusing point: The temperature at which metals or metallic alloys melt and become liquid.

Fusion piercing: A method of producing vertical blast holes by virtually burning holes in rock. Also known as piercing.

Fusion piercing drill: A machine designed to use the fusion-piercing mode of producing holes in rock.

—□—□—

Feather key

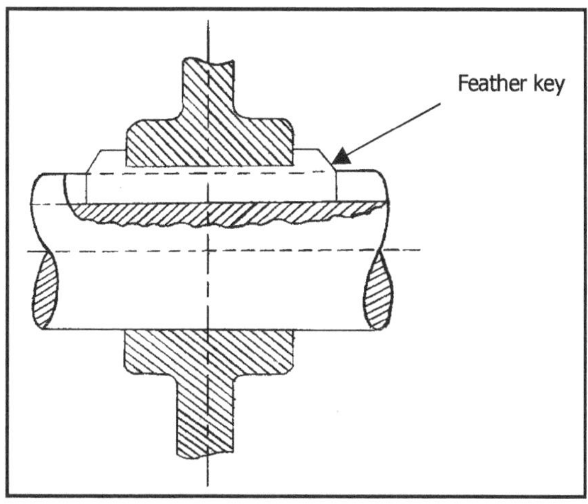

Fig. 1F Feather or sunk key

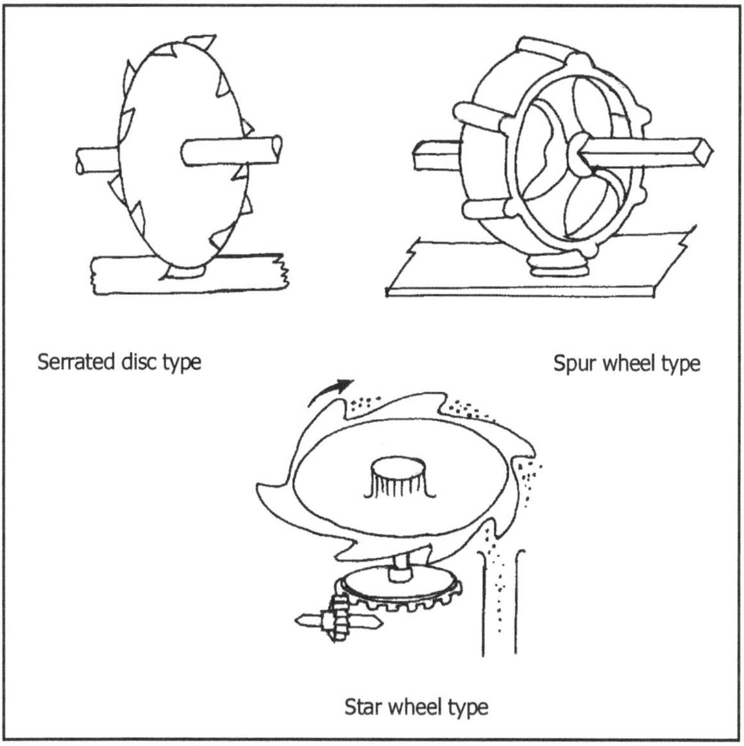

Serrated disc type
Spur wheel type
Star wheel type

Fig. 2F Feed mechanisms (fertilizer)

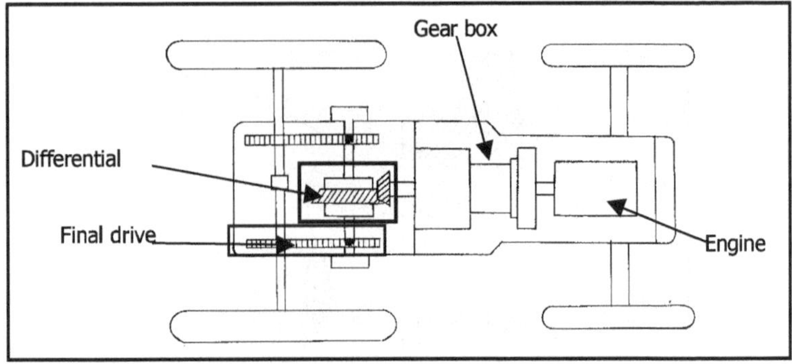

Fig. 3F Final drive

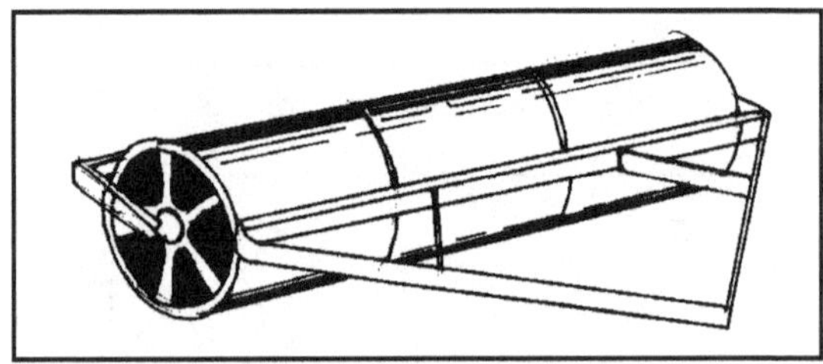

Fig. 4F Flat roller

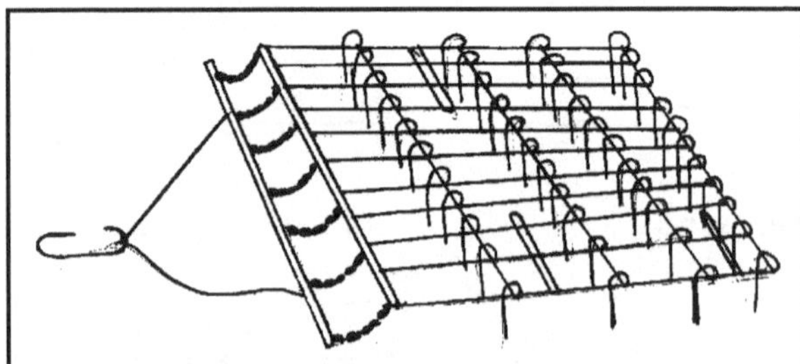

Fig. 5F Flexible tyne harrow

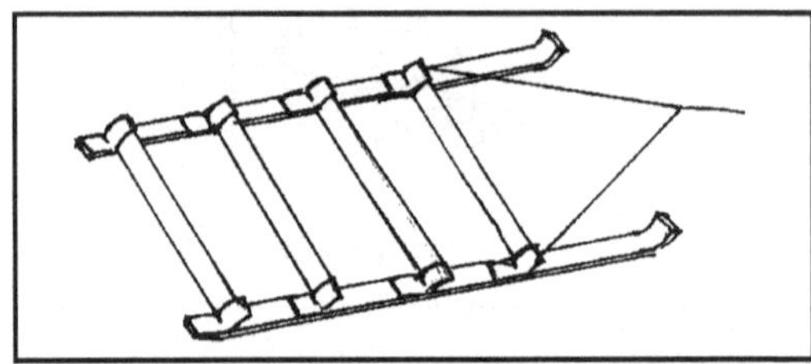

Fig. 6F Float

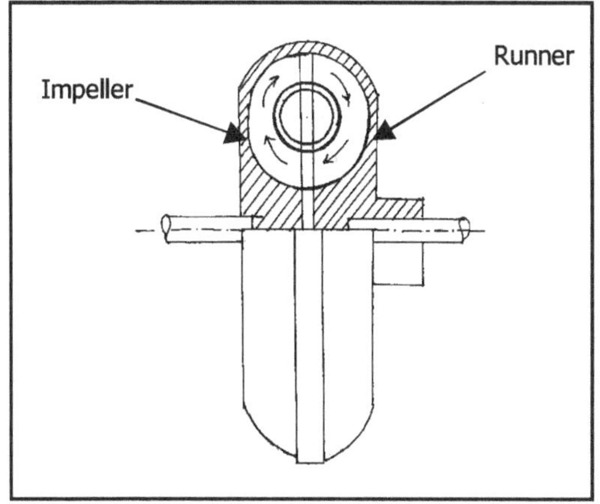

Fig. 7F Fluid coupling

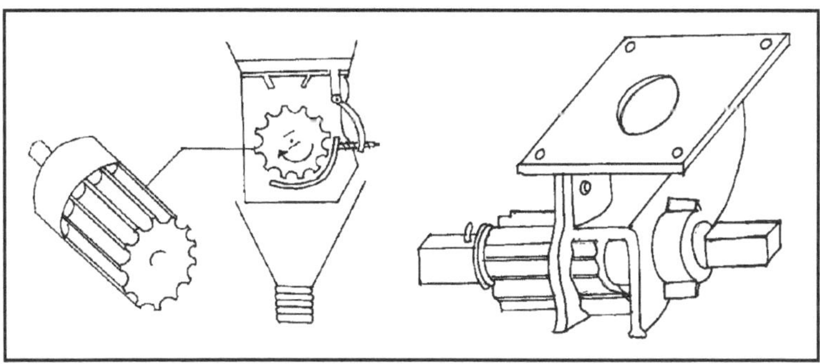

Fig. 8F Fluted feed roller

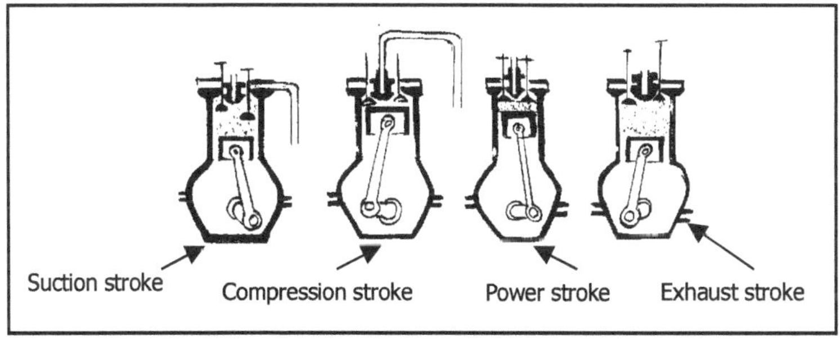

Fig. 9F Four stroke cycle

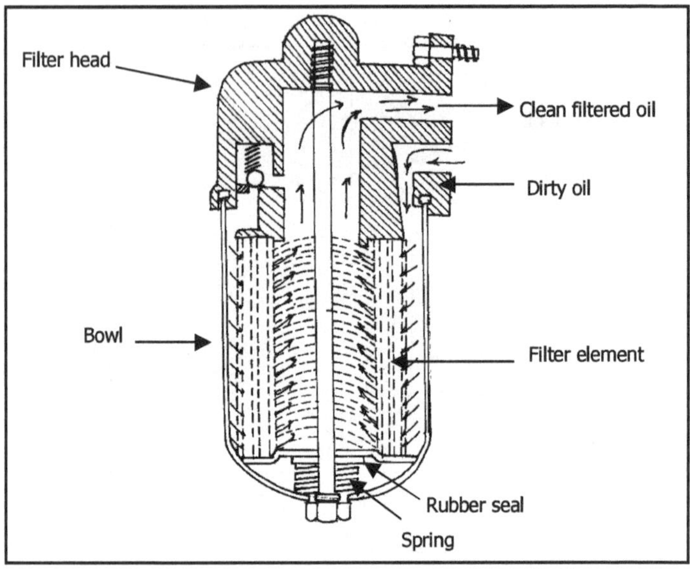

Fig. 10F Fuel filter

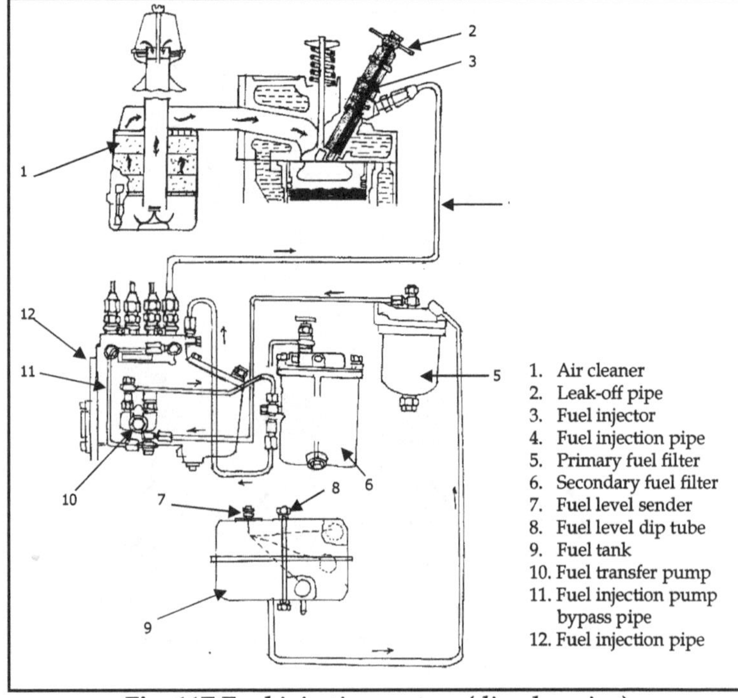

Fig. 11F Fuel injection system (diesel engine)

1
2
3
4
5
6
1. Fuel tank
2. Fuel tank filler cap
3. Filter gauze
4. Fuel filter and sediment trap
5. Fuel line
6. Carburetor

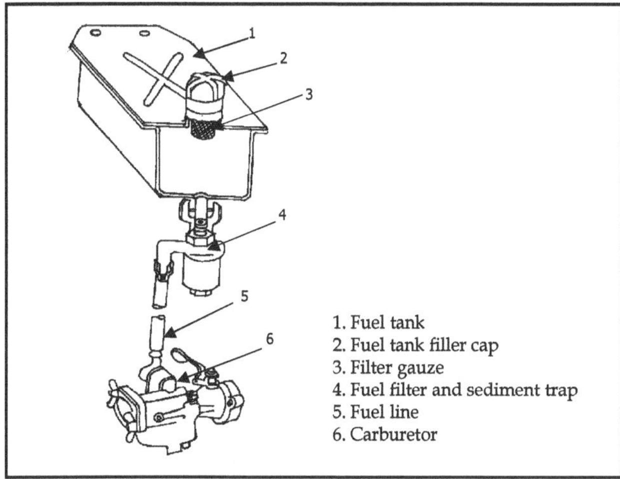

Fig. 12F Fuel system (petrol engine)

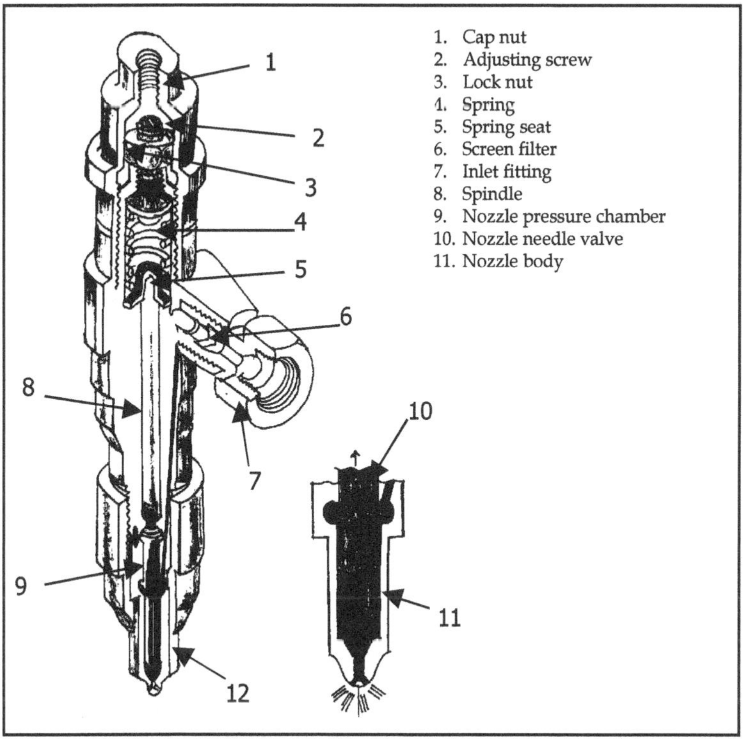

1. Cap nut
2. Adjusting screw
3. Lock nut
4. Spring
5. Spring seat
6. Screen filter
7. Inlet fitting
8. Spindle
9. Nozzle pressure chamber
10. Nozzle needle valve
11. Nozzle body
1
2
3
4
5
6
7
8
9
10
11
12

Fig. 13F Fuel injector

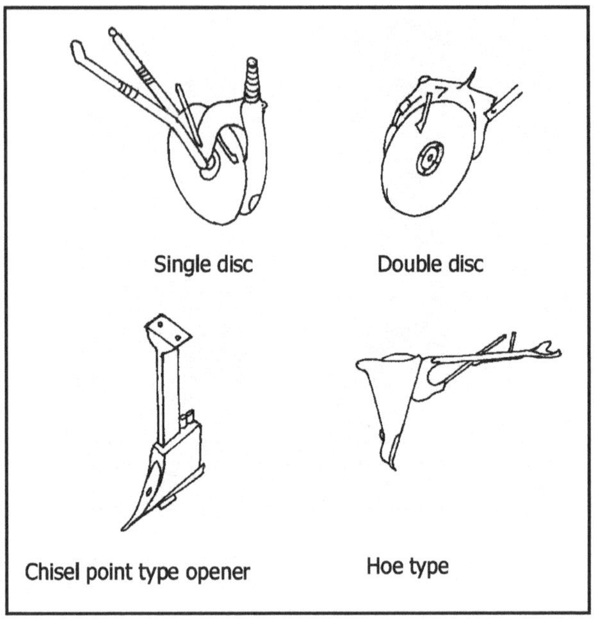

Fig. 14F furrow openers

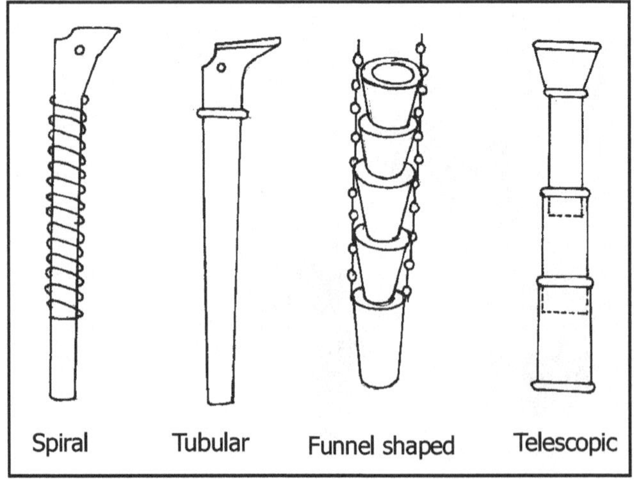

Fig. 15F Fertilizer/seed dropping units

Gadget: A slang word applied to any small, handy appliances or devices.

Gagging tape: A metal measuring tape used to determine the depth of liquid in a tank.

Gal.: The unit of acceleration in the centimeter-gram-second system, equal to 1 centimeter per second square; commonly used in geodetic measurement. Formerly known as Galileo, symbolized as Gal.

Galling: A characteristic of metals, which causes them to seize when, brought into intimate contact with each other. A material, which is subjected to galling, is one, which will seize or 'freeze' when brought into close contact with like material under pressure and no lubricant.

Gallows plough: An animal-drawn plough, the end of the beam rests freely on a bolster fixed to two front wheels, which is drawn through the front wheel assembly.

Galvanic corrosion: The accelerated corrosion of one metal when it is connected with a more noble metal, e.g. if steel and copper are connected together in seawater, the steel will suffer accelerated corrosion and the copper will be protected.

Galvanize: To coat iron with zinc. It is not usually an electrical process, but consists simply of dipping the iron in molten zinc.

Galvanized iron: Iron, which, after having undergone a cleansing process, has been dipped in a bath of molten zinc.

Galvanizing: The deposition of a zinc coating on iron.

Galvanometer: An instrument for detecting small currents or difference of potential. *See* Fig.1G.

Gandasa: An iron plate attached to a wooden frame and having cutting edge on one side and operated by hand.

Gang (machinery)**:** An assembly of concave discs mounted on common shaft with spools in between. *Refer* Fig. 9D.

Gang angle: The angle between the axis of the gang and line perpendicular to the direction of travel.

Gang axle (arbor axle) **:** A shaft on which a set of discs are mounted. *Refer* Fig. 9D.

Gang control lever: A lever, which operates the angling mechanism of disc harrow. *Refer* Fig. 9D.

Gang dies: Two or more punches and dies in one holder for making an equivalent number of openings in a bank with one stroke of the process.

Gang drill: A set of drills operated together in the same machine; used in rock drilling.

Gang drilling machine: A drill press equipped with several spindles so that a number of holes can be drilled at one time. It is particularly useful in quantity production work.

Gang mills: A series of milling cutters arranged on the same mandrel to increase production; may be used to machine two or more surfaces at one time.

Gang mower: An assembly of two or more ground driven cylinder mowers.

Gang plough: A riding plough with two or more bottoms.

Gang saw: A steel frame in which thin, parallel saws are arranged to operate simultaneously in cutting logs.

Gang tool: A tool holder with several cutting tools arranged so each tool cuts a little deeper than the one in advance of it. Gang tools are used on lathes and planers.

Gantt chart: In production planning and control, a type of bar chart depicting the work planned and done in relation to time; each division of space represents both a time interval and the amount of work to be done during that interval.

Gap lathe: An engine lathe with a sliding bed providing enough space for turning large-diameter work.

Gap (spark plug)**:** The space between the electrodes of a spark plug. *Refer* Fig.4S.

Gape length: Original length of the portion of a specimen measured for strain, length changes, and other characteristics.

Garden fork: A hand tool with handle and prongs placed in the line of handle or perpendicular to it used for loosening the soil. *See* Fig.2G.

Garden hatchet: An axe like tool with one side sharp edge having long handle used for cutting branches. *See* Fig.3G.

Garden rake: A hand tool used for collecting uprooted weeds, breaking clods and to some extent leveling of seeds bed in the garden. *See* Fig.5G.

Garden trowel: A transplanting trowel used for digging and lifting small plant. *See* Fig.6G.

Gas: A state of matter in which the matter has neither a definite shape nor a definite volume. Also an abbreviation for gasoline in North America.

Gas bearing: A journal or thrust bearing lubricated with gas. Also known a gas-lubricated bearing.

Gas brazing: A brazing process wherein the heat is obtained from a gas flame.

Gas compressor: A machine that increases the pressure of a gas or vapour by increasing the gas density and delivering the fluid against the connected system resistance.

Gas engine: An internal combustion engine designed to operate on a mixture of gas and air as a fuel.

Gas furnace: An enclosure in which a gaseous fuel is burned.

Gas hole: A cavity formed in a casting as a result of cavitations.

Gas injection: Injection of gaseous fuel into the cylinder of an internal combustion engine at the appropriate part of the cycle.

Gas manometer: A gage for determining the difference in pressure of two gases, usually by measuring the difference in height of liquid columns in the two sides of U- tube.

Gas mask: A device to protect the eyes and respiratory tract from noxious gases, vapors, and aerosols, by removing contamination with a filter and a bed of adsorbent material.

Gas meter: An instrument for measuring and recording the amount of gas flow through a pipe.

Gas power cycles: Are the cycles where the working fluid remains a gas throughout the entire cycle. Spark-ignition automobile engines, diesel engines, and conventional gas turbines are familiar examples of devices that operate on gas cycles.

Gas thermometer: A device to measure temperature by measuring pressure exerted by a definite amount of gas enclosed in a constant volume; the gas (preferably hydrogen or helium) is enclosed in a glass of fused quartz bulb connected to a mercury manometer. Also known as constant-volume of gas thermometer.

Gas turbine: A heat engine that converts energy of fuel into work by using compressed, hot gas as the working medium and that usually delivers its mechanical output through a rotating shaft. Also known as combustion turbine.

Gas turbine engine: A type of internal combustion engine in that the shaft is spun by the pressure of combustion gases flowing against curved turbine blades located around the shaft.

Gas vent: A pipe or hole that allows gas to pass off.

Gaseous fertilizer distributor: A machine for applying mineral fertilizer in gaseous form, which injects the fertilizer under pressure into the ground.

Gasholder: Gas storage container with vertically free top section that moves up or down to adjust to the volume of the gas held.

Gasket: A compressible insert, usually made of cork or metal or both, that is placed between two machined surfaces to provide a tight seal between them. *See* Fig.7G.

Gasohol: A blend of 90% gasoline and 10% ethanol used as an automotive fuel.

Gasoline: A volatile distillate from crude petroleum (blend of hydrocarbons), used principally as a fuel in internal combustion engines.

Gasoline engine: An internal combustion engine designed to operate on a mixture of gasoline and air as a fuel.

Gathering width: The distance between the centerline of the outermost divider pointes expressed in millimeters where adjustable divider is used, the maximum and minimum dimension shall be stated.

Gauge : An instrument or device for determining the size of parts. Gauges for different purposes are known by specific names.

Gauge pressure: A pressure read on a scale that ignores atmospheric pressure. Gauge pressure = absolute pressure - atmospheric pressure. Thus, the atmospheric pressure of 14.7 psi absolute is equivalent to 0 psi gauge (PSIG) the pressure above atmospheric pressure - as shown on dial of a pressure gauge in pounds per square inch.

Gauge stick: A graduated stick may be inserted in a tank or reservoir for determining the depth or quantity of the contents.

Gauge wheel: An auxiliary wheel of an implement to maintain uniform depth of working.

Gay Lussac law: The absolute pressure of a given mass of perfect gas varies directly as its absolute temperature, when the volume remains constant.

Gear: A very general term applied to toothed wheels. *See* Fig.8G.

Gear case: A housing or metal box within which gears operate. Usually the gear case is filled with lubricant.

Gear cutter: Circular cutter of hardened steel whose section is that of the tooth spaces, which they are intended to cut.

Gear forming: A method of gear cutting in which desired tooth shape is produced by a tool whose cutting profile matches the tooth form.

Gear generating: A method of gear cutting in which the tooth is produced by the conjugate or total cutting action of the tool plus the rotation of the work piece.

Gear grinding: A gear cutting method in which gears are shaped by formed grinding wheels and by primarily a finishing operation.

Gear level: To arrange gears also that the driven part and driving part turn at the same speed.

Gear loading: The power transmitted or the contact force per unit length of a gear.

Gear pump: A rotary pump (volumetric pump) in which two meshing gear wheels contra rotate so that the fluid is entrained on one side and discharged on the other.

Gear ratio: 1. The relation between the speed of driving and driven gears. 2. The number of revolutions made by a driving gear as compared to the number of revolutions made by a driven gear of different size. For example, if one gear makes three revolutions while the other gear makes one revolution, the gear ratio would be 3 to 1.

Gear shaper: A machine that makes gear teeth by means of a reciprocating cutter that rotates slowly with the work.

Gear teeth: Projections on the circumference or face of a wheel, which engage with complementary projections on another wheel to transmit force and motion. *See* Fig.9G.

Gear train: An arrangement of two or more gears connecting driving and driven parts.

Gear up: To arrange gears so that the driven pat rotates faster than the driving part.

Geared chuck: A form of universal chuck.

Geared head: A headstock equipped with back gear.

Geared pump: A pump, which is driven by an engine through the use of gearing.

Gearing: The term has the same general meaning as gear, but is applied more specifically to gearwheels.

Gearing chain: A continuous chain used to transmit motion form one toothed wheel, or sprocket, to another.

Gearshift: A device for engaging and disengaging gears.

Gearshift lever: A lever by means of which the change speed gears are shifted.

General-purpose mouldboard: A mouldboard having a medium curvature lying between stubble and sod. Its surface is slightly convex having gradual slope. *See* Fig.10G.

Generator: A device that converts mechanical energy to electrical energy. It is a machine used to produce current electricity by cutting lines of force with a conductor. Also called as a dynamo.

Generator efficiency: The ratio of the electrical power output to the mechanical power input.

Generator regulator: The device used to control the pressure (voltage) and the amount (amperage) of generator current output with reference to the varying speeds of the governor.

Germination: Growth of the radical and plumule in seed, when the later absorbs adequate water and is placed in a normal environment of temperature and humidity.

Gib: 1. That portion of a gib and cotter arranged used in the strap end of a connecting rod to keep it from spreading. It is a flat piece of steel with hook ends. 2. A thin piece of steel used as an adjusting strip to bring about a perfect sliding fit in machine parts. Adjustment is usually secured by the pressure of setscrews against the gib.

Gibbs free energy: The amount of energy available to do useful work resulting from a chemical reaction. When hydrogen and oxygen combine to form water, the useful energy is equivalent to: 217 BTU/mole water (: 229 kJ/ mole water).

Gibbs-Dalton law: An extension of Dalton's law of additive pressures, states that under the ideal-gas approximation, the properties of a gas in a mixture are not influenced by the presence of other gases, and each gas component in the mixture behaves as if it exists alone at the mixture temperature and mixture volume.

Gibheaded key: A key having a prong or offset standing at right angles with the thicker end to facilitate drawing it back. *See* Fig.11G.

Gill: 1. A unit of volume used in the United State for the measurement of liquid substances equal to ¼ U.S. liquid pint or to 1.1829411825 x 10^{-4} cubic meter.

Girdle: A traction aid with a series of stakes, which are connected together by chains or an articulated frame.

Gland: The small bearing which closes the mouth of a stuffing box and takes the wear of the piston. *See* Fig.12G.

Glass: A hard, brittle substances made by melting sand or silica with lime, potash, soda, or lead oxide.

Glasscutter: Any device used for cutting glass to size; usually a diamond or a small rotary wheel set in a handle. *See* Fig.13G.

Glazed lining: A brake lining which has become coated and smooth as a result of heat on the lining itself or of foreign substances which may have become deposited upon or imbedded in it.

Go or no-go gauge: Often a double end gauge with certain allowable tolerance between the ends. One end fits nicely the part being gauged; the other end is too small for an outside diameter or too large for an inside diameter.

Gold: The most malleable and ductile of all metals. Symbol, Au.

Governor: A mechanical device, designed to control the speed of an engine within specified limit, used on tractor or stationary engines. Governing system is broadly classified as 1. Hit and miss system and 2. Throttle system. See hit and miss system and throttle system.

Governor regulation (R) : It is the variation in the engine speed between full load and no load condition. It is usually expressed as percentage of rated speed and given as :

$$R \text{ (in percent)} = \frac{N_1 - N_2}{(N_1 - N_2)/2} \times 100$$

Gpm: Gallons per minute.

Gps: Gallons per second.

Grab: A device used for hauling or hoisting.

Grabbing clutch: A friction clutch, which does not engage gradually but takes hold suddenly causing the car to be jerked. It is a rough performing clutch.

Grader: A high-bodied, wheeled vehicle with a leveling black mounted between the front and rear wheels; used for fine grading relatively loose and level earth.

Grading: The process of sorting material into different lots confirming to certain predetermined standards.

Graduate: To divide into regular steps or grades, as a scale.

Graduation: The method or system of dividing a graduated scale.

Grafting knife: A knife to cut the wood from scion and to insert the same in the stock. *See* Fig.14G.

Grain auger: The device, which carries the grain-to-grain elevator.

Grain cleaner: A machine to remove foreign matter from grain mass. This may be termed on the basis of grain, such as paddy and seed.

Grain discharge auger: The device which carries the grain from the tank to discharge elevator.

Grain discharge elevator: A device to discharge the grain from the tank to the grain auger.

Grain elevator: The device, which carries the grain from grain auger to grain tank or bin.

Grain grader: A machine used for grading the grain.

Grain pan: The pan for collecting the clean grain after being passed through cleaning sieve for conveying to grain auger.

Grain processing: The process of upgrading grains for improving their marketability, storability and suitability for human consumption.

Grain separator: A machine to remove impurities from grain or other seeds and to sort them into two or more fractions.

Grain tank: A tank to hold the grain after having received it from grain elevator.

Gram force: A unit of force in the centimeter gram second gravitational system, equal to the gravitational force on a one gram mass at specified location. Abbreviated as gF.

Grams per mile (Gpm): A measure for the weight of pollutants emitted into the atmosphere with the vehicle exhaust gases. Antipollution laws set maximum limits for each exhaust pollutant in grams per mile.

Granule applicator: An appliance for applying materials in the form of granules.

Granule hopper: Container for holding granules.

Granule metering mechanism: Part of granule applicator, which controls the flow of granules at desired application rate.

Granule nozzle: A device for directing granules in airflow.

Graphite: Contains more than 90% carbon, having a peculiar silvery luster, and is used in the manufacture of lead pencils, as a lubricant, and in foundry work as a mold coating.

Grapple skidder: Skidder that uses a suspended grapple or bottom opening jaws to assemble and hold its load.

Grass catcher: A part or combination of parts, which provide a means for collecting grass clippings or debris.

Grass shear: A scissor like tool used for cutting the grass.

Grass sword: A thin metal strip with a handle having one side cutting edge used for cutting the grass normally in bending position.

Grating: An arrangement of bars used to cover an opening. Also used for forming platforms in engine rooms, fire escapes, etc.

Gravimetric analysis: One way to describe the composition of a mixture that is accomplished by specifying the mass of each component.

Gravitational acceleration: Having value as 9.807 m/s^2 at sea level and varies by less than 1 percent up to 30,000 m. Therefore, g can be assumed to be constant at 9.81 m/s^2.

Gravitational constant: The constant of proportionality in Newton's law of gravitation force between any two particles times the square of the distance between them, divided by the product of their masses. Also known as constant of gravitation.

Gravity: The force which draws all bodies toward the center of the earth or to its surface.

Gravity feed drill: A seeder in which the metering of seeds is done through gravity and positive metering is not possible.

Gravity feed fertilizer distributor: A gravity-feed fertilizer distributor fitted with a mechanical distribution mechanism, which provides a width of spread greater than the width of the hopper.

Gravity feed seeder: A seeder in which the metering of seeds is done through gravity and positive metering is not possible.

Gravity irrigation: Irrigation in which water is not pumped but flows by gravity including sprinkler system.

Greenhouse effect: The tendency for the atmosphere to hold in infrared radiation, or heat, because of increased levels of carbon dioxide (CO_2).

Grind: To polish a surface by means of an abrasive wheel.

Grip: A device for grasping or holding.

Gross load: The sum of payload and unladen mass of the trailer.

Gross weight: The weight of a vehicle or container when it is loaded with goods; abbreviated gr. wt.

Ground clearance of power tiller: The height of the lowest point of the power tiller chassis from a firm horizontal supporting surface.

Ground clearance of tractor: The height of the lowest point of the tractor chassis from a firm horizontal supporting surface, the tractor being ballasted as used for drawbar test.

Ground wheel-driven: Operated by a wheel in contact with the ground.

Grounded system (electrical) **:** The single wire system, which makes use of the metal of the automobile frame and body for one of the two conductors in an electrical circuit.

Groundnut decorticator: A machine to separate groundnut kernels from pods.

Groundnut digger: A machine to expose the pods of groundnut from the soil to facilitate their picking. *See* Fig.15G.

Groundnut digger shaker: A groundnut digger with a mechanism for separating soil from the pods.

Groundnut thresher: Equipment used for threshing the groundnut crop.

Grouser height: Vertical distance from the track shoe face to tip of the grouser.

Grouser: A transverse strake, incorporated in or attachable to a track, to assist adhesion between the track and the ground.

Grouser width: Overall width of the grouser.

Growler: A device used to determine if any of the coils in an armature are grounded or short.

Guard bar: A steel bar to carry guards wearing plates and knife guide.

Guard: Any protective device attached to or used in connection with a machine to reduce liability of injury to the operator.

Gudgeon pin: *See* wrist pin or piston pin.

Guide bar cover: Removable device for covering the guide bar and chain when the saw is not being used.

Guide bar: The parts that supports and guides the saw chain.

Guide bearings: Bearings, which consist of a channel or groove in which part slide; e.g., crosshead, bearings.

Guide pin: A pin used to line up a tool that bears direction work.

Guide pulley : Pulley to lead a driving belt or rope in a new direction or to keep it from leaving it's desired direction.

Guide rail: A rail placed on the inside of the main rail of railroad track to guide the wheel flanges; used principally on curves and bridges.

Gunnel (plough): The vertical face of the share, which rubs against the furrow wall. *Refer* Fig. 1S.

Gunter's chain: A chain 66 feet (20, 1168 meters) long, consisting of 100 steel links, each 7,92 inches (20.1168 centimeters) long, jointed by rings, which is used as the unit of length for surveying public lands in the United States. Also known as 'chain'.

Gyroscope: An instrument that maintains as angular reference direction by virtue of a rapidly spinning, heavy mass; all applications of the gyroscope depends on a special from of Newton's second law, which states that a massive, rapidly spinning body rigidly resists being disturbed and tends to react to a disturbing torque by processing (rotating slowly) in a direction at right angles to the direction of torque.

Gyroscopic couple: The turning moment, which opposes any change of the inclination of the axis of rotation of a gyroscope.

Gyroscopic: The branch of mechanics concerned with gyroscopes and their use in stabilization and control of ships, aircraft, projectiles, and other objects.

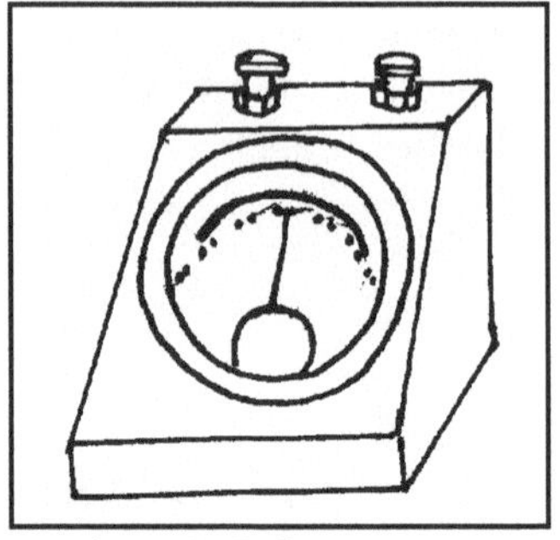

Fig. 1G Galvanometer

Fig. 2G Garden fork

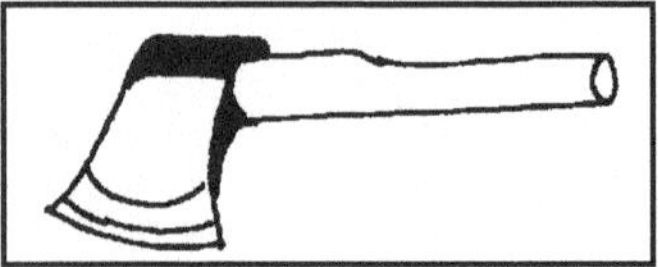

Fig. 3G Garden hatchet

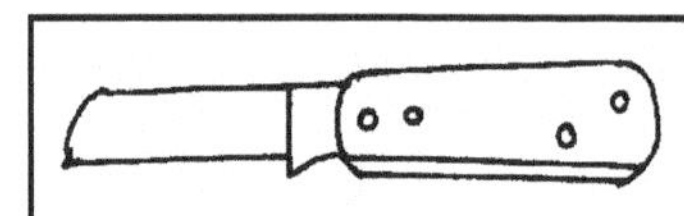

Fig. 4G Garden knife

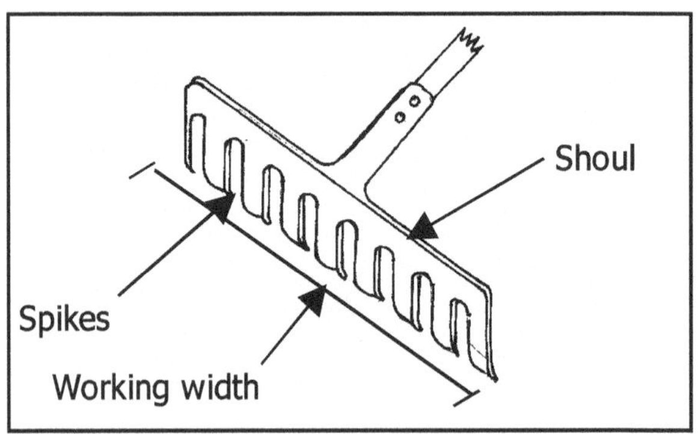

Fig. 5G Garden rake

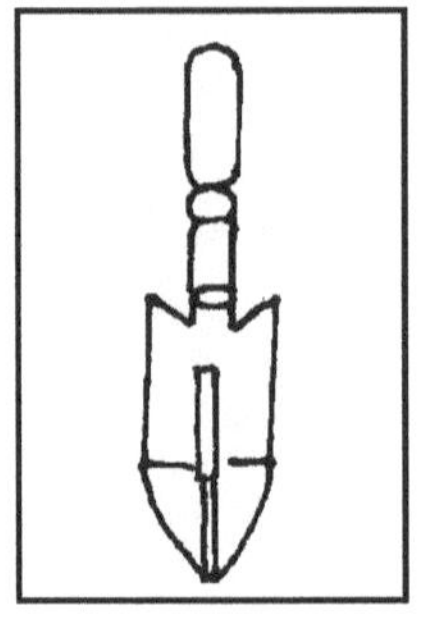

Fig. 6G Garden trowel

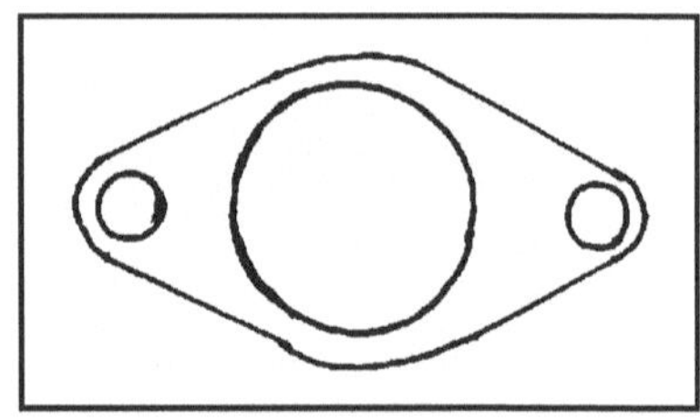

Fig. 7G Gasket

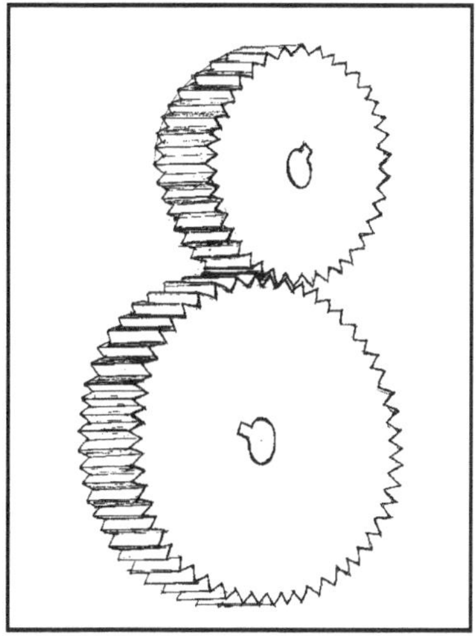

Fig. 8G Gear

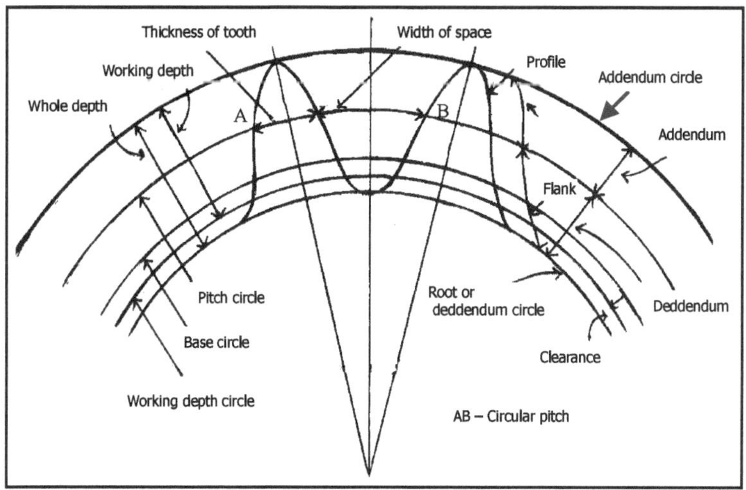

Fig. 9G Gear tooth parts

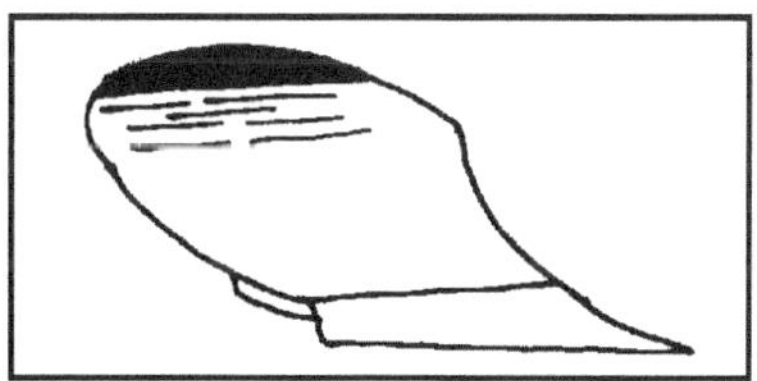

Fig. 10G General purpose mouldboard

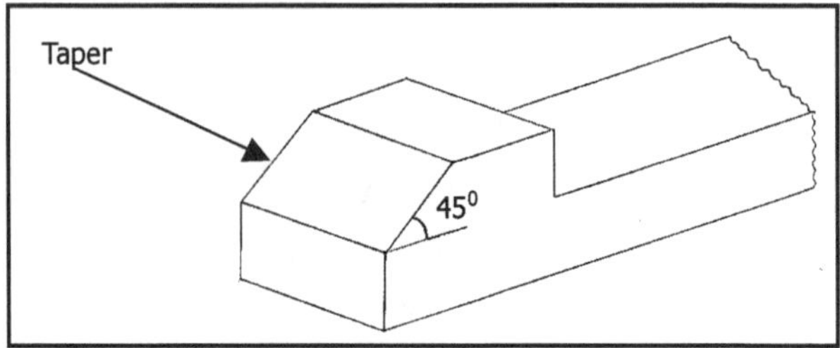

Fig. 11G Gib headed key

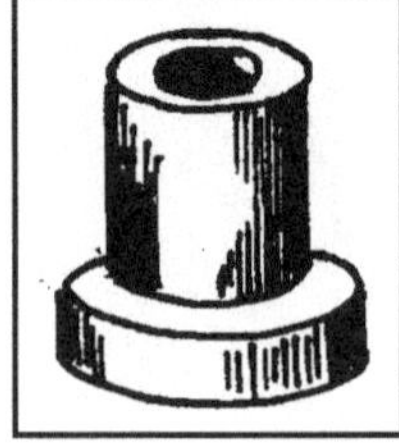

Fig. 12G Gland

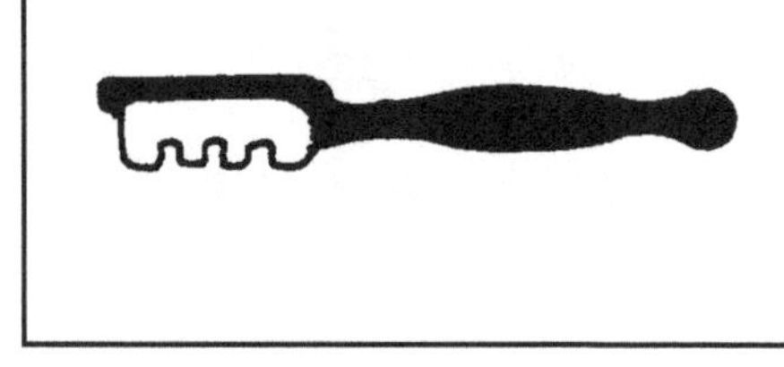

Fig. 13G Glass cutter

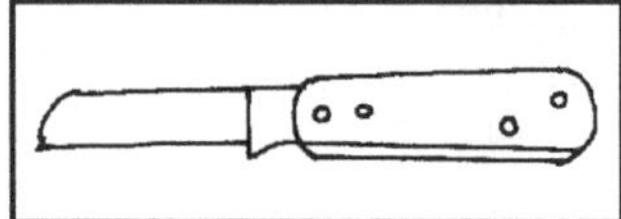

Fig. 14G Grafting knife

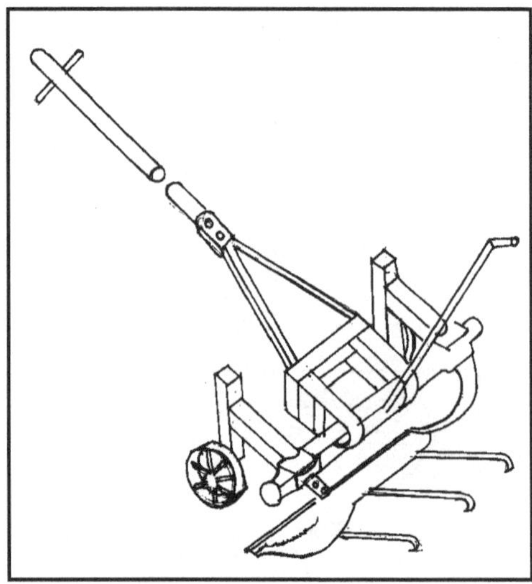

Fig. 15G Groundnut digger (Animal drawn)

Half round file: A file, which is flat on one side and curved on the other. The amount of convexity never equals a semicircle.

Half section: In mechanical drawing, a sectional view that terminates at the centreline, showing an external view on one side of the centerline and on the other side an interior view.

Hammer: An instrument or tool used for striking blows in metalworking, driving nails, etc. Hammers are of various kinds, each bearing a name according to the purpose it has to serve.

Hammer mill: 1. A type of impact mill or crusher in which materials are reduced in size by hammers revolving rapidly in a vertical plane within a steel casing. Also known as beater mill. 2. A grinding machine, which pulverizes feed and other products by several rows of thin, hammers revolving at high speed.

Hand brake: A brake operated by hand, used principally as a parking brake. When applied, it remains set until released by hand. *Refer* Fig. 2E.

Hand cultivator: A hand tool with prongs and handle used for collecting the weeds and breaking the clods in the garden.

Hand dibbler: A peg type tool used for making holes in soil for placing the seeds. *Refer* Fig. 3D.

Hand drill: A drilling machine operated by hand.

Hand hoe: A hand-operating hoe. *See* Fig.1H.

Hand operated: Operated by hand.

Hand punch: A hand held device for punching holes in paper or cards.

Hand tools: Tools that are guided and operated by hand. *See* Fig.2H (eg. Hand Sprayer)

Hand-feed tuber planter: A tuber planter, the planting element of which is fed by hand.

Handrail: A narrow rail to be grasped by a person for support.

Handsaw: An ordinary one handled saw, either rip or cross-cut, used by woodworkers.

Hardening: The process of heating the steel to a temperature above critical point, holding it at this temperature for a considerable period and then quenching in water, oil or molten sand bath. Due to hardening, material becomes resistant to wear. Its strength, elasticity, ductility and toughness gets increased.

Harmonic motion: A periodic motion which is a sinusoidal function of time. Harmonic that is, motion along a line given by the equation $x=a \cos (kt + q)$, where t is the time parameter, and a, k, and q are constants. Also known as harmonic vibration or simple harmonic motion (SHM).

Harrow: An implement that cuts the soil to a shallow depth for smoothening and pulverizing it, to cut weeds and to mix materials with the soil.

Harrowing: A secondary tillage operation, which pulverizes, smoothens, and packs the soil in seedbed preparation and weeds control.

Harvester: A self-propeller multi-function machine, which combines felling with other processing functions.

Harvesting: The operation of detaching picking or cutting the crop from the undesired portion of the same rooted to the ground.

Hay loader: A machine to pick up and load hay from windrows.

Hay rake: The machine to collect dried cut crops into transverses windrows.

Hazard: 1.An existing or potential condition that can result in an accident. 2. Any risk to which a worker is subjected as a direct result (in whole or in part) of his being employed.

Hazardous: A substance or circumstance that may cause injury or damage by reason of being explosive, poisonous, corrosive, oxidizing, or otherwise harmful.

HC: Hydrocarbon; used to represent emissions from an internal combustion engine.

Head rice: Rice grains of size ¾ and above in length of whole rice grain shall be called head rice.

Head stock: 1. The device on a lathe for carrying the revolving spindle. 2. The movable head of certain measuring machines. 3. The device on a cylindrical grinding machine for rotating the work.

Header: The portion of the combine comprising the mechanism for gathering, cutting, stripping or picking the crop and deliver it to the cylinder.

Header width: The distance between the sides sheets of the feed table measured immediately above the tips of the knife expressed in meters. Where the feed table is offset from the centre line of the machine, the amount of offset and weather it is to left or right shall be stated.

Headless set screw: A setscrew, which, instead of having a head, has a slot to permit adjustment by a screwdriver.

Heat: A form of energy that is released by the burning of fuel. In an engine, heat energy is converted to mechanical energy.

Heat (thermodynamics): The form of energy that is transferred between two systems (or a system and its surroundings) by virtue of a temperature difference.

Heat engine: A device that convert heat into work. Heat engines are differing considerably from one another, but all can be characterized by the following: 1.They receive heat from a high-temperature source (solar energy, oil furnace, nuclear reactor, etc.). 2.They convert part of this heat to work (usually in the form of a rotating shaft). 3. They reject the remaining waste heat to a low-temperature sink (the atmosphere, rivers, etc.). 4. They operate on a cycle.

Heat exchangers: Device where two moving fluid streams exchange heat without mixing. The simplest form of a heat exchanger is a double-tube (also called tube-and-shell) heat exchanger composed of two concentric pipes of different diameters. One fluid flows in the inner pipe, and the other in the annular space between the two pipes. Heat is transferred from the hot fluid to the cold one through the wall separating them. Sometimes the inner tube makes a couple of turns inside the shell to increase the heat transfer area, and thus the rate of heat transfer.

Heat of compression: An increase in temperature brought about by the compression of a gas.

Heat pump: A cyclic device which transfers heat from a cooler reservoir to a hotter one, expending mechanical energy in the process, especially when the main purpose is to heat the hot reservoir rather than refrigerate the cold one.

Heat pump coefficient of performance: is the efficiency of a heat pump, denoted by COP_{HP}, and expressed as desired output (Q_H) divided by required input ($_{Wnet.\ in}$), i.e. $COP_{HP} = Q_H/_{Wnet.\ in}$.

Heat reservoir: The thermal energy reservoir since it can supply or absorb energy in the form of heat.

Heat sink: Heat reservoir that absorbs energy in the form of heat.

Heat source: Heat reservoir that supplies energy in the form of heat.

Heat-driven system: Refrigeration system, whose energy input is based on heat transfer from an external source. Absorption refrigeration system is often classified as heat-driven system.

Heat treatment: The careful controlled heating and cooling of steel to bring it to its highest efficiency.

Heating value of a fuel: The amount of heat released when a specified amount of fuel (usually a unit mass) at room temperature is completely burned and the combustion products are cooled to the room temperature. In other words, the heating value of a fuel is equal to the absolute value of the enthalpy of combustion of the fuel.

Hedge shear: A scissor life tool used for pruning of hedge and twigs.

Heel of landside: The rear underside of the landside, which slides along the furrow sole.

Height gage: A gage used to measure heights by either a micrometer or a vernier scale.

Height of power tiller: The distance between the firm horizontal supporting surface and horizontal plane touching uppermost part of the power tiller with the engine in horizontal position.

Height of tractor: The distance between the firm horizontal supporting surface and horizontal plane touching uppermost part of the tractor.

Helical conveyor: A conveyor for the transport of bulk materials, which consists of, a horizontal shaft with helical paddles or ribbons rotating inside a stationary tube.

Helical gear: Gears with teeth cut at an angle other than right angle across the face. Often incorrectly called spiral gear. It may be used to transmit power between (1) parallel shafts, (2) shafts at right angles and not intersecting, (3) shafts inclined at any angle and not intersecting. It gives greater strength and smoother operation, but develops considerable end thrust. *See* Fig.3H.

Helical spring: A compression-type spring shaped like the frustum of a cone.

Helix: A curve, as would be obtained by winding a thread around a cylinder in such a manner that there would be a uniform amount of advance with each revolution.

Helix angle: The angle made by the helix of the thread at the pitch diameter with a plane perpendicular to the axis.

Henry's law: The mole fraction of a weakly soluble gas in the liquid is equal to the partial pressure of the gas outside the liquid divided by Henry's constant.

Herringbone gear: A gear in which the teeth slope both ways from the center line of the gear face, as would be the case if two spiral gears, one left hand one right hand were fastened together; used for heavy work on mining machinery, etc.

Hexagon: A plane figure having six sides and six angles. All sides of a regular hexagon are equal. The sum of six internal angles equal to 720 deg.

Hexagon nut: The ordinary six-sided form of nut.

Hexagonal head bolt: A standard wrench heads bolt with a hexagonal head.

Hg: Chemical symbol for mercury.

HICE: Hydrogen Internal Combustion Engines.

High carbon steel: A rather general term applied to steels of more than 0.50 per cent carbon, of good tempering qualities, and suitable for cutting tools.

High resistance point (electricity)**:** any point in an electrical circuit which blocks the full flow of electrical current under normal voltage; may be caused by rust or other corrosion.

High volume (spray) **:** Spray volume more than 560 liters per hectare.

Higher heating value (HMV)**:** The heating value of the fuel when the water in the combustion gases is completely condensed and thus the heat of vaporization is also recovered. Efficiencies of furnaces are based on higher heating values.

High-pressure cylinder: Cylinders used for storing high-pressure gas with a service pressure of 1000 psi or greater.

High-tension armature type magneto: A magnet that carries a secondary winding over a primary winding on the armature to produce two waves of high voltage current for each revolution of the armature.

Hill divider (thresher)**:** The projection provided on steeped grain bed, chaffer, and cleaning sieve to prevent material from sliding to one side specially when working on slope.

Hill dropping: The process of placing seeds in small groups at regular intervals along straight parallel furrow and covering the seeds with soil so as to get assurance that there will be at least one plant from each spot where the group of seed are placed.

Hillside tractor: A wheeled tractor equipped with manual or automatic means of compensating for sloping ground.

Hip pad: Strap or pad of leather, plastic or other suitable material fastened either to the saws or to the harness, to cushion the operator from impact and to reduce transmission of vibration.

Hit and miss system (governor): In this system, the frequency of explosions or power strokes of an engine is regulated by controlling the opening and closing of exhaust valve through the linkage. As engine speed exceeds the rated speed, exhaust valve and suction valve gets open and close respectively, which prevents fresh charge to enter into the engine cylinder and hence power stroke is missed. *See* Fig.4H.

Hitch point: An articulated connection between link and implement for geometrical purpose, the hitch point is the center of the articulated connection between a link and the implement.

Hob: 1. A master cutter, for cutting worm wheels and spur gears. 2. A master model in hardened steel used to sink the shape of mold into a soft steel block. 3. A bracket in a fir-place on which a kettle may be hung.

Hobbing: The process of cutting the teeth of worm wheels, threads of dies, or chasers, with a hob or master tap.

Hoe: A hand tool with blade (s), tine (s) and sometime with disc (s) attached to a frame used for inter- cultivation of row crops. *See* Fig.1H and Fig 1H1.

Hoe and rake combined: A hand tool having one side rake and opposite side hoe.

Hoe-type furrow opener: A furrow opener consisting of a single or a double pointed shovel fastened to the lower part of the boot.

Hollow cone nozzle: A cone nozzle in which the formation of an air core within the orifice and the swirl chamber due to a higher rotational velocity results in the production of hollow cone of the liquid.

Hollow punch: A hardened steel punch with a hammer for cutting holes in metal, cardboard, fabric, etc.

Homogeneous: Made up of similar parts of the same quality throughout.

Homogenizer: A machine that blends or emulsifies a substance by forcing it through fine openings against a hard surface.

Hone: An abrasive tool for correcting small irregularities of differences in diameter in a cylinder.

Hooke's law: This law states that within the elastic limit the deformation produced is proportional to the stress.

Hopper: A box or receiver used for the purpose of feeding material to the machine.

Horizontal: In the direction of or parallel to the horizon; on a level.

Horizontal auger: A rotary drill, usually powered by a gasoline engine, for making horizontal blasting holes in quarries and opencast pits.

Horizontal boiler: A boiler, the longitudinal axis of which is horizontal.

Horizontal boring machine: A machine tool having a horizontal spindle adjustable both vertically and longitudinally. The worktable is also adjustable and may be rotated, making the machine adaptable to a wide range of work.

Horizontal clearance or horizontal clearance (MB plough): The maximum clearance between landside and a horizontal plane touching point of share at its gunnel side and heel of landside. It is also known as horizontal suction. *See* Fig.5H.

Horizontal clevis: A device to effect lateral adjustment of the plough relative to the line of pull.

Horizontal drilling machine: A drilling machine in which the drill bit extend in a horizontal direction.

Horizontal lathe: A horizontally mounted lathe with which longitudinal and radial movements are applied to a work piece that rotates.

Horizontal milling machine: A milling machine with horizontal spindle and cutter arbor. The table, which can be raised and lowered to suit work conditions, is capable of horizontal feed.

Horizontal rotary mower: A mower with high-speed knives rotating in the horizontal plane.

Horse power (HP): the energy required to lift 75 kg by a height of one meter in one second against gravity. One horsepower equals 746 Watt, or 0.746 kW.

Hose clamp: A clamp for making a tight joint between a hose and some less flexible object over which the end of the hose fits.

Hose coupling: A device, generally a union, for joining together the ends of hose lengths.

Hot plug: A spark plug with long porcelain runs warmer than one with short porcelain; therefore, it is often refereed to as a hot plug.

Hot spot: Refers to a comparatively thin section of area of the wall between the inlet and exhaust manifold of an engine, the purpose being to allow the hot exhaust gases to heat the comparatively cool incoming mixture. Also used too designate local areas of the cooling system, which have attained above average temperatures.

Housing: A term of very general application usually referring to a body casting, a main part, a container, a cover, or support for other parts.

Humidifier: An apparatus for supplying moisture to the air and for maintaining desired humidity conditions.

Humidifying: The process of adding moisture to atmospheric air.

Hummeter plate: A plate attached below the feed of the concave for removing awns from the grain.

Hybrid electric vehicle (HEV): A vehicle that is powered by both an electric drive system and a second source of power, such as an internal combustion engine, referred to as the alternative power unit (APU).

Hydraulic agitation: Agitation of the spray mixture by using an auxiliary pump flow or a partial flow of the main pump.

Hydraulic agitator: A device used for hydraulic agitation.

Hydraulic brake system: The system of automotive brakes, which is operated by means of pressure within the fluid system of the brake. The fluid system consists primarily of master cylinder tubing, and wheel cylinders. *See* Fig.6H and 6H1.

Hydraulic brake: A system of operating mechanical brakes by power supplied by hydraulic pressure. When pressure is supplied to the pedal a piston is forced into a master cylinder, and the fluid contained in it (a mixture of denatured alcohol and castor oil) is distributed through copper tubes or flexible tubing to various parts of the system.

Hydraulic circuit: A circuit whose operation is analogous to that of an electric circuit except that electric currents are replaced by currents of water fluids, as in a hydraulic control.

Hydraulic cylinder: The cylindrical chamber of a positive displacement pump. *Refer* Fig. 6H.

Hydraulic drill: A rotary drill powered by hydrodynamic means and used to make short-firing holes in the rock, or to make a well hole.

Hydraulic drive: A mechanism transmitting motion from one shaft to another, the velocity ratio of the shafts being controlled by hydrostatic or hydrodynamic means.

Hydraulic dynamometer: *See* Fig.7H.

Hydraulic injector: A device using the velocity of a jet of liquid to produce a vacuum in a suction pipe for the purpose of filling a tank.

Hydraulic machine: A machine powered by a motor activated by the confined flow of a stream of liquid, such as oil or water under pressure.

Hydraulic motor: A motor activated by water or other liquid under pressure.

Hydraulic nozzle: An atomizing device in which fluid pressure is converted into fluid velocity.

Hydraulic power lift: A member driven by a tractor's power lift to raise, hold and lower mounted or semi mounted equipments by hydraulic means.

Hydraulic power system: A power transmission system comprising machinery and auxiliary components which function to generate, transmits, control, and utilize hydraulic energy.

Hydraulic pressure spray: Spray, which is projected to the target by its own pressure.

Hydraulic ram: A device for forcing running water to a higher level by using the kinetic energy of flow; the flow of water in the supply pipeline is periodically stopped so that a small portion of water is lifted by the velocity head of a larger portion. Also known as hydraulic pump.

Hydraulic scale: An industrials scale in which the load applied to the load cell piston is converted to hydraulic pressure.

Hydraulic separation: Mechanical classification using a hydraulic classifier.

Hydraulic spray nozzle: A type of nozzle used for hydraulic spraying.

Hydraulic sprayer: 1. A machine that sprays large quantities of insecticide of fungicide on crops. 2. A sprayer used for hydraulic spraying. *See* Fig.8H.

Hydraulic spraying: Spraying performed by the use of hydraulic energy.

Hydraulic stacker: A tiering machine whose carriage is raised or lowered by a hydraulic cylinder.

Hydrocarbon (HC): An organic compound containing only carbon and hydrogen, usually derived from fossil fuels such as petroleum, natural gas, and coal: an agent in the formation of photochemical smog.

Hydrocarbon fuels: are the most familiar fuels and consist primarily of hydrogen and carbon. They are denoted by the general formula C_nH_m. Hydrocarbon fuels exist in all phases, some examples being coal, gasoline, and natural gas.

Hydroelectric plant: A facility at which electric energy is produced by hydroelectric generators. Also known as hydroelectric power station.

Hydrogen (H_2) : The simplest and lightest element in the universe, which exists as a gas except at low cryogenic temperatures. Hydrogen gas is colorless, odorless and highly flammable gas when mixed with oxygen over a wide range of concentrations. Hydrogen forms water when combusted, or when otherwise joined with air, as within a fuel cell. Hydrogen molecules in which both protons have the same spin are known as 'orthohydrogen'. And those in which the protons have opposite spins are known as 'Para hydrogen'.

Hydrometer: *See* Fig.9H.

Hydrostatic brakes: A brake system, containing fluid, which is hermetically sealed. Pressure on the foot pedal distorts the fluid reservoir and pressure is distributed equally to all four wheels. At each wheel a brake fluid chamber expands outward, forcing six brake shoes against the brake drum.

Hypoid gears: A type of spiral bevel gears, which permit the location of the pinion above or below the center of the gear with which it meshes. Hypoid gears are now largely used for the rear axle drive in all passenger automobiles.

Hypoid generator: A gear-cutting machine for making hypoid gears.

Hypoid: An abbreviation of hyperboloid. The term is applied to a special type of spiral bevel gear tooth.

Hypothesis: an assumption as a basis for investigation or reasoning.

Hysteresis clutch: A clutch in which torque is produced by attraction between induced poles in a magnetized iron ring and the control field.

Hythane: A commercial gas product that contains 20% hydrogen and 80% natural gas.

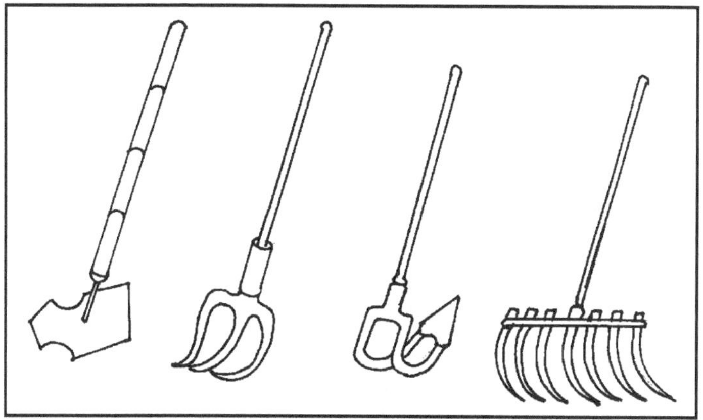

Fig. 1H Hand hoes

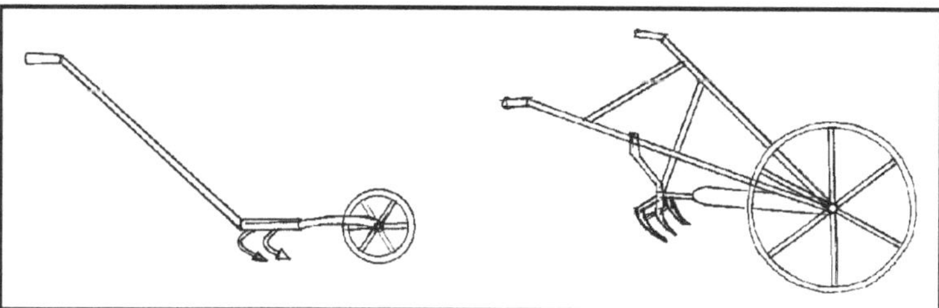

Fig. 1H1 Wheel hoe

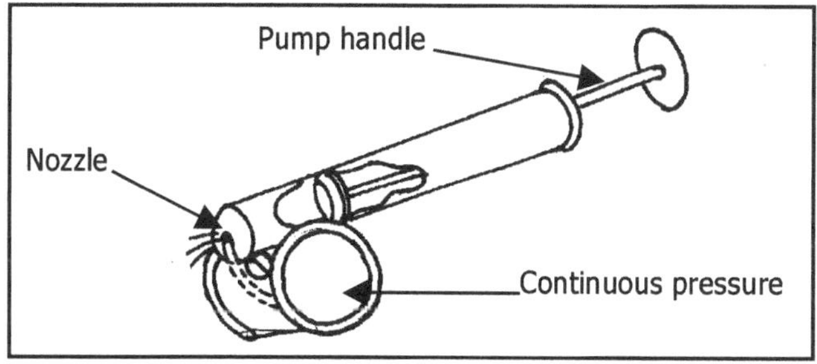

Fig. 2H Hand sprayer

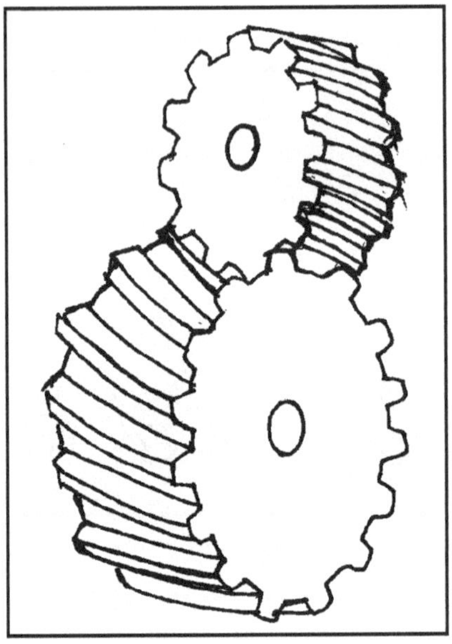

Fig. 3H Helical gear

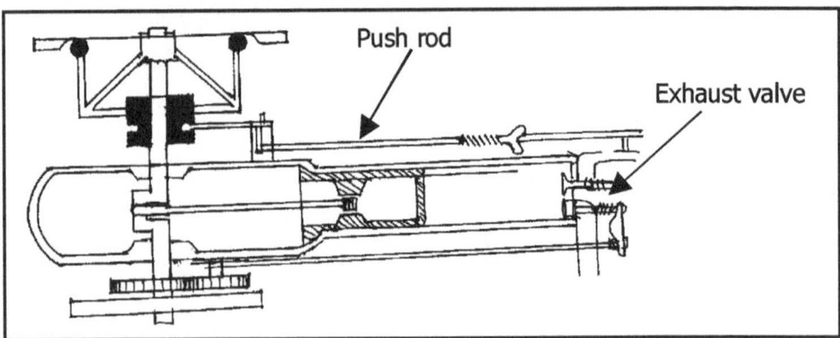

Fig. 4H Hit and miss governing system

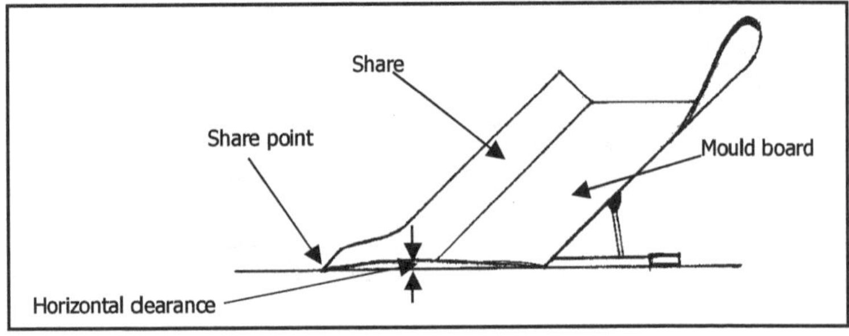

Fig. 5H Horizontal clearance

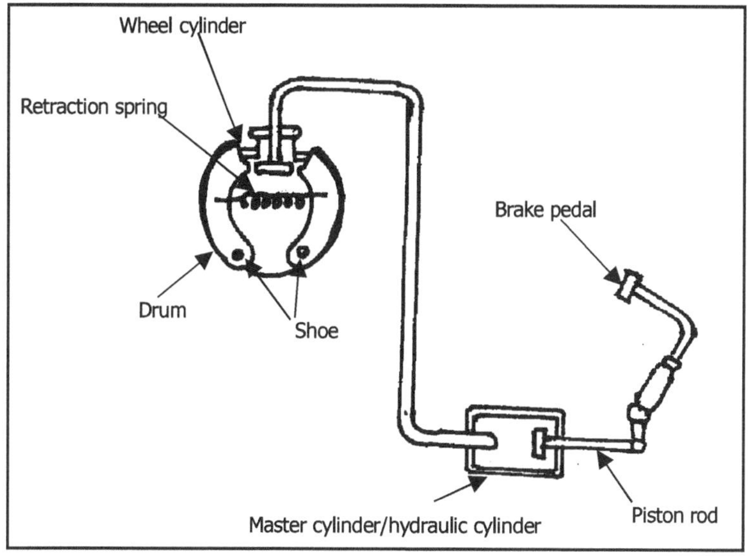

Fig. 6H Hydraulic brake system

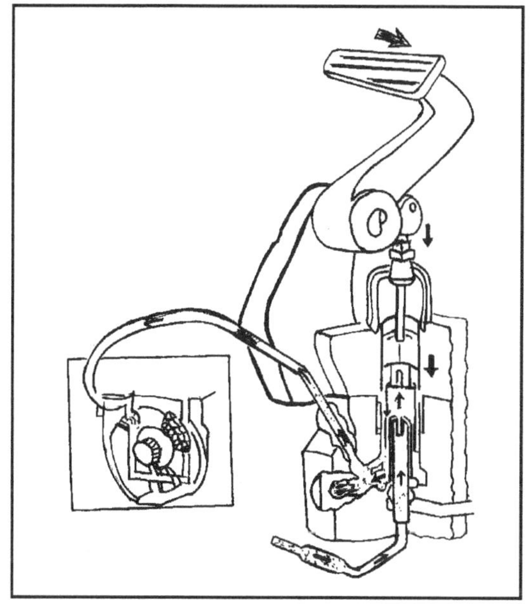

Fig. 6H1 Hydraulic brake system

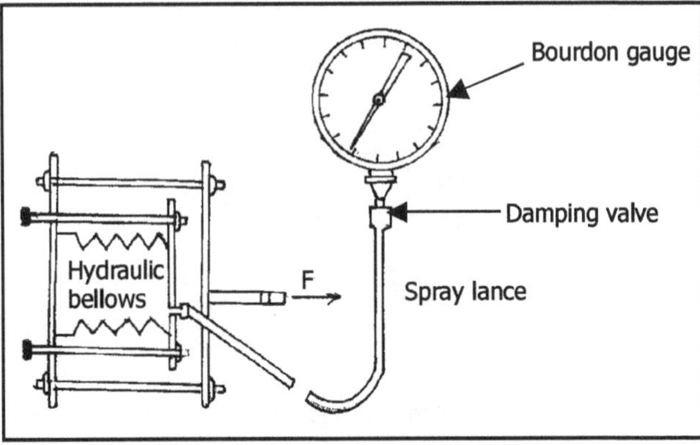

Fig. 7H Hydraulic dynamometer

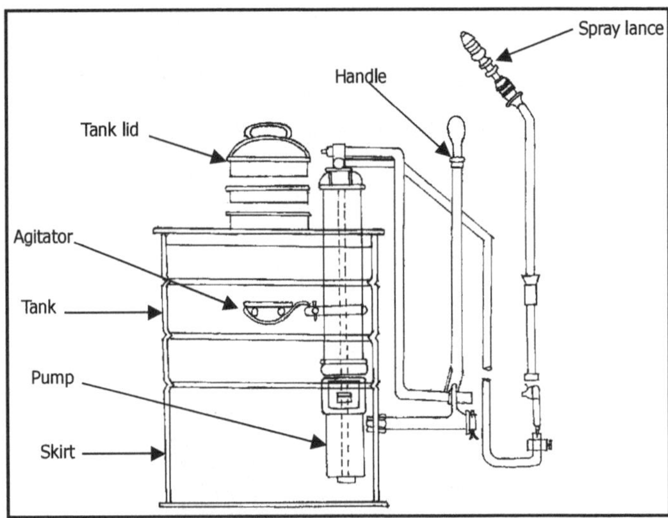

Fig. 8H Hydraulic sprayer

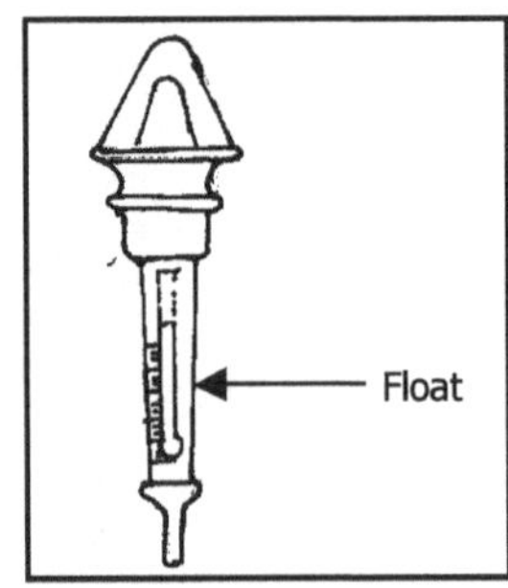

Fig. 9H Hydrometer

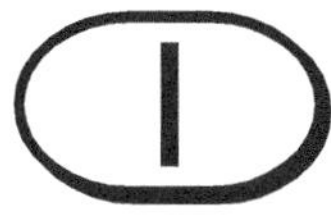

I beam: A steel beam shaped like the letter I, used in structural work.

I.H.P. (Indicated Horse Power): 'Pure' horse power as measured in the combustion chamber before friction and other loses are subtracted.

IC Engine (ICE): Internal Combustion Engine.

Ideal compression-compression refrigeration cycle: completely vaporizes the refrigerant before it is compressed and expands the refrigerant with a throttling device, such as an expansion valve or capillary tube. The compression-compression refrigeration cycle is the most widely used cycle for refrigerators, air-conditioning systems, and heat pumps. It consists of four processes:1-2 Isentropic compression in a compressor; 2-3 Constant-pressure heat rejection in a condenser; 3-4 Throttling in an expansion device; 4-1 Constant-pressure heat absorption in an evaporator

Ideal cycle: is an actual cycle stripped of all the internal irreversibilities and complexities. The ideal cycle resembles the actual cycle closely but is made up totally of internally reversible processes.

Ideal gas: Is a gas that obeys the ideal-gas equation of state.

Ideal gas law (third gas law): Approximates the behavior of many gases under conditions close to ordinary atmospheric temperatures and pressures.

Ideal gas temperature scale: Is a temperature scale that turns out to be identical to the Kelvin scale. The temperatures on this scale are measured using a constant-volume gas thermometer, which is basically a rigid vessel filled with a gas, usually hydrogen or helium, at low pressure.

Ideal gear: A gear situated between a driving gear and a driven gear to transfer motion, without any change of direction or of gear ratio.

Idle speed : Engine speed when the accelerator is fully released, and there is no load on the engine.

Idler pulley: A pulley used to guide and tighten the belt or chain of a conveyor system. *See* Fig. 1I

Idler wheel: 1. A wheel used to transmit motion or to guide and support something. 2. A roller with a rubber surface used to transfer power by frictional means in a sound recording or sound reproducing system.

Idling: Refers to the slow running of the engine when the vehicle is not in motion.

Idling jet: The jet, which controls the amount of gasoline needed for operating the engine at idling speed. Always taken off above the butterfly valve.

Idling system: A system to obtain adequate metering forces at low airspeeds and small throttle openings in an automobile carburetor in the idling position.

Ignition: 1.The act of setting on fire the compressed fuel charge in the combustion chamber before friction and other losses are subtracted. 2. The action of the spark in starting the burning of the compressed air: fuel mixture in the combustion chamber.

Ignition by hot tube or bulb: In this spark ignition, tube or bulb like projection, closed at one end and the other end is opened in the combustion space is used to ignite the charge. This tube or bulb is externally heated. During the compression stroke some charge enters in the hot tube or bulb and is ignited by the contact with hot inner surface. This method of ignition is not commonly used.

Ignition coil: An induction coil so designed and built as to provide high-tension spark for ignition of the compressed charge. It steps up low voltage to a very high voltage (about 20,000 volts). *See* Fig. 2I

Ignition energy: The amount of external energy that must be applied in order to ignite a combustible fuel mixture.

Ignition lag: In the internal combustion engine, the time interval between the passage of the spark and the inflammation of the air fuel mixture. Also known as Ignition delay.

Ignition spark: The high-tension spark or arc induced in the secondary winding of the ignition coil and passed to the spark plugs where it explodes the gas in the engine cylinders.

Ignition switch: An electrical control, which completes or breaks the circuit to ignition coil either by wiping contacts or push pull, or toggle fingers.

Ignition system: The arrangements of different components, for igniting charge in engine cylinder, at proper time in internal combustion engine is called as ignition system. There are four systems of igniting fuel as 1. Spark ignition, 2. Compression ignition and 3. Ignition by hot tube or bulb. Refer spark ignition, compression ignition and ignition by hot tube or bulb.

Ignition temperature: The minimum temperature to which a fuel must be brought to start the combustion.

Ignition timer: A mechanical device, usually a cam by which time of sparking is set to provide effective combustion.

Immediate surroundings: The portion of the surroundings that is affected by the process.

Impact test: The testing of materials for resistance to shock.

Impeller: The rotating member in a pump, usually consisting of a number of vanes, which puts in motion the medium through which it travels. *See* Fig.3I.

Impulse: As applied to impulse starter or impulse fuel pump, it refers to momentary acceleration given to speed of engine, shaft, or armature to produce starting, increase of vacuum, or advance of timing position. Applied to auto engine and aviation engine.

Inboard drive: The bendix starting motor pinion, which moves inward towards the starting motor when the switch is closed.

Incomplete combustion: A combustion process in which the combustion products contain any unburned fuel or components such as C, H_2, CO, or OH.

Incomplete lubrication: Lubrication that takes place when the load on the rubbing surfaces is carried partly by fluid viscous film and partly by areas of boundary lubrication; friction is intermediate between that of fluid and boundary lubrication.

Incompressible substances: Substances having negligible variation with pressure.

Increase of entropy principle or second law of thermodynamics: The entropy of an isolated system during a process always increases or, in the limiting case of a reversible process, remains constant. In other words, the entropy of an isolated system never decreases.

Increment: The amount by which a varying quantity increases between two of its stages.

Indefinite: Not precise, unsettled, uncertain.

Indexing (machining) : Dividing the circle into regular spaces, for purposes of milling, fluting, and gear cutting.

Induction coil: It consists of a numbers of soft iron rods or strips bound together by an insulating material, which forms its center.

Induction motor: A type of AC electric motor, in which the rotor has no direct connection to an external source of electricity.

Inert gas: A gaseous component in a chemical reaction that does not react chemically with the other components.

Inertia: A property which tends to keep a motionless body at rest or also tends to keep a moving body in motion; effort is thus required to start mass moving or to retard or stop it, once it is in motion.

Information system: It is a system, which collects and process data and disseminates information.

Ingot: A mass of metal, which after being purified, is cast from such metals as gold, copper, tin, etc. Ingots are usually rectangular in section, and bear the imprint of the manufacturer.

Injection pump: A pump that forces a measured amount of fuel through a fuel line and atomizing nozzle in the combustion chamber of an internal combustion engine. *Refer* Fig. 11F.

Injector (diesel) **:** An assembly, which receives a metered fuel from another source at relatively low pressure, then is actuated to inject it into a cylinder or chamber at high pressure. *Refer* Fig.13F.

Inlet port. The passageway through which the fuel charge is fed to the cylinder.

Inline engine: An engine in which the cylinders are arranged one behind the other.

Inline linkage (steering) **:** A power steering linkage, which has the control valve and actuator combined in a single assembly.

Inside calipers: A calipers with the points at the end of the legs turned outward instead of inward, so that it may be used for gauging inside diameters.

Inside thread: An internal thread, a thread cut on an inside diameter as to receive a bolt.

Inspect: To examine a component or system for its function.

Installation: Placing in position or ' setting up' of machinery or power plant equipment etc.

Installment: 1. A delivery in part. 2. A partial payment.

Insulation: A material that prevents the transfer of electricity or heat.

Insulator: Any material used to prevent the flow of electrical current, an electrical non-conductor.

Intake manifold: The passage, which permits a fluid or gas to enter a chamber and seals against exit. *Refer* Fig. 4A

Intake stroke: The charge (air or mixture of air and fuel) admission phase.

Intake valve: An inlet, through which the charge (air or air-fuel mixture) is drawn into the cylinder. *Refer* Fig. 4A

Integral: Built into, as part of the whole.

Intensifier: A device often used in place of a hydraulic accumulator for converting a low pressure into a higher pressure.

Intensive properties: Are those that are independent of the size of a system, such as temperature, pressure, and density.

Interchangeable: Refers to similar parts so accurate in manufacture that they can be substituted one for another.

Interlocking: Fastening together as by a dovetail joint.

Intermittent firing: Cyclic firing whereby fuel and air is burned in a furnace for frequent short time periods.

Intermittent gear: Gear where the teeth are not continuous, but have plain surface between.

Internal brake: A friction brake in which an internal shoe follows the inner surface of the rotating brake drum, wedging, itself between the drum and the point at which it is anchored; used in motor vehicles.

Internal combustion engine : An engine in which the fuel is burned inside the engine cylinder itself, rather than in a separate device.

Internal furnace (boiler) **:** A boiler furnace having a firebox within a water-cooled heating surface.

Internal gear drive: Any drive in which an internal (annular) gear meshes with a spur pinion in order to increase or decrease speed ratio.

Internal gear: Where spur wheels or pinions engage with teeth set on the internal circumference of a ring, the gear is called "internal". The reverse of spur gear.

Internal thread: A thread cut inside a piece of material as, for example, in a nut.

Internally reversible process: The process in which no irreversibilities occur within the boundaries of the system during the process. During an internally reversible process, a system proceeds through a series of equilibrium states, and when the process is reversed, the system passes through exactly the same equilibrium states while returning to its initial state.

Inventory: The itemized list of goods or stock on hand in a business.

Inversion temperature: The temperature at a point where a constant-enthalpy line intersects the inversion line.

Inverted engine: An engine in which the cylinders are below the crankshaft.

Invoice: Itemized list of purchases and charges, sent to a buyer.

Involute teeth: Most commonly used form of gear teeth, development being based on involute curve.

Involute: If a piece of string wound about a cylinder at a certain point should be unwound and kept taut during the unwinding, the end of the string would describe the curve known as an involute.

Iron: The metallic element which plays the most important part in the industrial world. It is obtained from ores in combination with other substances. Iron is marked as cast, wrought, malleable, and steel.

Irregular: Departing from or being out of the usual or proper form or order.

Irregular curve: A draftsmen's tool used for drawing curves, which are not arcs of circles. Also called French curve or Universal curve.

Irreversibility: 1.The factors, causing a process to be irreversible. It include friction, unrestrained expansion, mixing of two gases, heat transfer across a finite temperature difference, electric resistance, inelastic deformation of solids, and chemical reactions. 2.The difference between the reversible work W_{rev} and the useful work W_u due to the irreversibilities present during the process. Irreversibility can be viewed as the wasted work potential or the lost opportunity to do work.

Irreversible process: The processes which, once having taken place in a system, cannot spontaneously reverse itself and restore the system to its initial state.

Isentropic: Having constant entropy; at constant entropy.

Isentropic compression: Compression, which occurs without any changes in entropy.

Isentropic efficiency of a compressor: The ratio of the work input required to raise the pressure of a gas to a specified value in an isentropic manner to the actual work input.

Isentropic efficiency of a turbine: The ratio of the actual work output of the turbine to the work output that would be achieved if the process between the inlet state and the exit pressure were isentropic.

Isentropic process: An internally reversible and adiabatic process. In such a process the entropy remains constant.

Isentropic stagnation state: The stagnation state when the stagnation process is reversible as well as adiabatic (i.e. isentropic). The entropy of a fluid remains constant during an isentropic stagnation process.

Iso: Prefix is often used to designate a process for which a particular property remains constant.

Isobaric process: A process during which the pressure remains constant.

Isochoric (or isometric) process: A process during which the specific volume remains constant.

Isolated system: A closed system in which energy is not allowed to cross the boundary.

Isothermal compressibility: Relates how volume changes when pressure changes when temperature is held constant.

Isothermal efficiency of a compressor: The ratio of the work input to a compressor for the reversible isothermal case and the work input to a compressor for the actual case.

Isothermal process: A process during which the temperature remains constant.

Fig. 1I Idler pulley

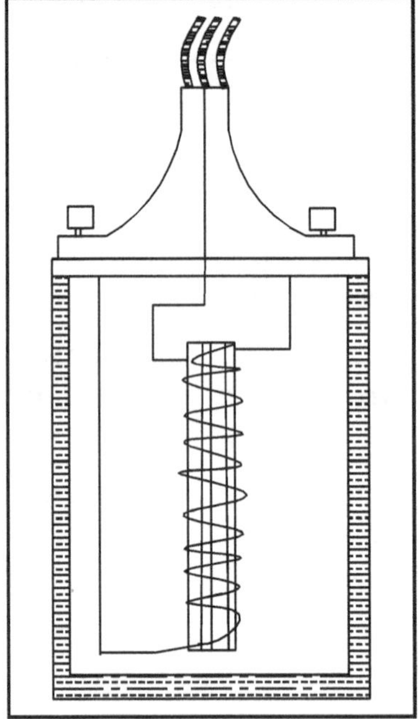

Fig. 2I Ignition coil

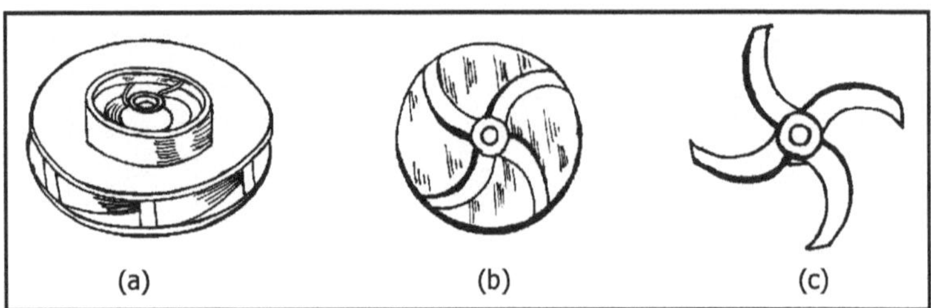

Fig. 3I Impeller types

Jack, electric: A form of metallic spring contact, connection being made by inserting a plug, which is attached to a cord. Commonly used on telephone switchboards and radios.

Jack, hydraulic: Device for raising weight or exerting pressure by pumping oil or other liquid under a piston or ram.

Jack: A mechanical device used for lifting heavy loads through short distances with a minimum expenditure of manual power.

Jacket: An outer casing, as around a boiler or tank, to preserve heat or cold, or around a motor cylinder, permitting a flow of water to prevent overheating.

Jacking up: The raising up of masses of machinery and heavy structures by means of jacks.

Jackscrew: Small screw jacks for leveling work in jigs.

Jackshaft: A countershaft, especially when used as an auxiliary shaft between two other shafts.

Jet compressor: A device, utilizing an actuating nozzle and a combining tube, for the pumping of a compressible fluid.

Jig: A device, which holds and locates a piece of work and guides the tools, which operate upon it.

Jig, drill: A device for holding work while drilling, having bushings through which the drill is guided so that the holes are correctly located in the piece. Milling and planning jigs (commonly called "fixtures") hold the work while operate upon it.

Joint, universal: A shaft connection, which allows freedom in any direction and still conveys a positive motion. Most of them can transmit power through any angle up to 45 deg.

Jointer: It is a device which resembles a miniature plough bottom, used to cut and turn over a small ribbon like furrow slice directly ahead of the main bottom of the plough.

Joule (J): Metric unit of energy, work, heat or torque. Equal to the Newton meter (Nm). Named for English physicist James Prescott Joule (1818:1889) who pioneered the measurement and understanding of heat.

Joule's law: The change of internal energy of a perfect gas is directly proportional to the change in temperature.

Joule Thomson coefficient: A measure of the change in temperature with pressure during a constant-enthalpy process.

Journal: A part or support within which a shaft operates.

Journal bearing: A cylindrical bearing, which supports a rotating cylindrical shaft.

Journal friction: Friction of the axle in a journal bearing arising mainly from viscous sliding friction between journal and lubricant.

Jump spark: A spark produced when electricity jumps across an air gap between two conductors to which electrical cables are attached.

Junction: Place of union; point of meeting; joint.

Junction box: A terminal plate or bar fitted with screws to which the terminal ends of electrical cables are attached. A box like cover protects entire arrangement.

Junker's engine: A double opposed-piston, two-cycle internal combustion engine with intake and exhaust ports at opposite ends of the cylinder.

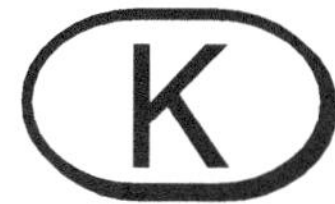

Kelvin scale: The thermodynamic temperature scale in the SI and is named after Lord Kelvin (1824-1907). The temperature unit on this scale is the Kelvin, which is designated by K (not °K; the degree symbol was officially dropped from Kelvin in 1967). The lowest temperature on the Kelvin scale is 0 K.

Kelvin unit: magnitude was established at the International Conference on Weights and Measures in 1954. The triple point of water (the state at which all three phases of water exist in equilibrium) was assigned the value 273.16°K (0.01°C). The magnitude of a Kelvin is defined as 1/273.16 of the temperature interval between absolute zero and the triple-point temperature of water. The magnitudes of temperature units on the Kelvin and Celsius scales are identical (1 K, 1°C). The temperatures on these two scales differ by a constant 273.16.

Kelvin-Planck statement of the second law of thermodynamics: It is impossible for any device that operates on a cycle to receive heat from a single reservoir and produces a net amount of work. This statement can also be expressed as no heat engine can have a thermal efficiency of 100 percent, or as for a power plant to operate, the working fluid must exchange heat with the environment as well as the furnace.

Key: 1. A small block inserted between the shaft and hub 2. A wedge -shaped strip of iron or steel used for preventing wheels from slipping around (preventing circumferential movement) upon their axles. *Refer* Fig. 11G.

Keyway: A groove in which a key is placed for the purpose of binding something, as a crank, gear, or pulley, on a shaft.

Kiln: An oven or furnace for baking, burning, or drying. i.e. a kiln for drying lumber, a kiln for burning bricks, a kiln for burning lime etc.

Kilo: Prefix in metric system meaning one thousand; also shortened form of kilogram.

Kilo Pascal (kpa): Metric unit of pressure. One psi equals 6.9 kpa. One inch of mercury (Hg) equals 3.38 kpa. Related units are the Pascal (Pa) at 1000 per kpa, the bar at 100 kpa and the mega Pascal at 1000 kpa. See also Pascal.

Kilocalorie: A unit of heat energy equal to 1000 calories. Abbreviated as kcal.

Kilogram (kg): Metric unit of weight or mass, equal to approximately 2.2 lb. 1 kg =2.2 lb = 1000 g.

Kilo joule: (1 kJ) is 1000 joules.

Kilometer (km): Metric unit of length, equal to 0.62 miles (1 mile =1.6 km). Related units: 1 kilometer= 1000 metre, 1 metre = 10 decimetre, 1 metre = 100 centimetre and 1 metre = 1000 millimetre)

Kilopascal (kpa) : The unit of pressure equal to 1000 Pascal or 1000 N/m^2.

Kilowatt (kW): A unit of power equal to about 1.34 hp, or 1000 watts. The Watt is named for James Watt, Scottish engineer (1736:1819), a pioneer in steam engine design.

Kinetic: Active motion, as opposed to potential.

Kinetic energy (KE): Energy that a system possesses as a result of its motion relative to some reference frame. When all parts of a system having mass m move with the same velocity v, the kinetic energy is expressed as $KE = m V^2/2$.

Kingpin: A steel pin or shaft, hardened and ground to size, used in fastening and supporting the steering knuckle to the axle beam, permitting free movement right and left of the front wheels.

Kingpin inclination: The side-to-side inclination of the king pin. This serves the same purpose as chamber. Some manufactures combine camber with a slight king pin inclination for front wheel alignment. *Refer* Fig. 5C.

Knock, or engine knock: Audible noise occurring in the engine because of auto ignition or the premature ignition of the fuel, caused by excessive clearance between moving parts in an engine. In petrol engine, a knock can be overcome by increasing the octane number of the fuel, retarding ignition timing or by reducing engine load. In diesel engines, knock is caused by excessive pressure within the combustion chamber and abnormal fuel combustion. In diesel engines it is avoided by the use of higher Cetane fuels. Knocking may be resulting from worn or improperly fitted pistons, piston (wrist) pins, connecting rod, or main bearings.

Knockoff: 1.The automatic stopping of a machine when it is operating improperly. 2. The device that causes automatic stopping.

Knuckle joint: One in which an eye at the end of a rod is engaged by the forked end of a second rod, the two being connected with a pin.

Knuckle thread: A thread which is half round at both root and crest, can be cast in a mold, and is rarely used except on rough work. *See* Fig. 1K.

Knurling: Impressing a design into a metallic surface usually by means of small hard rollers that carry the corresponding design on their surfaces.

—□—□—

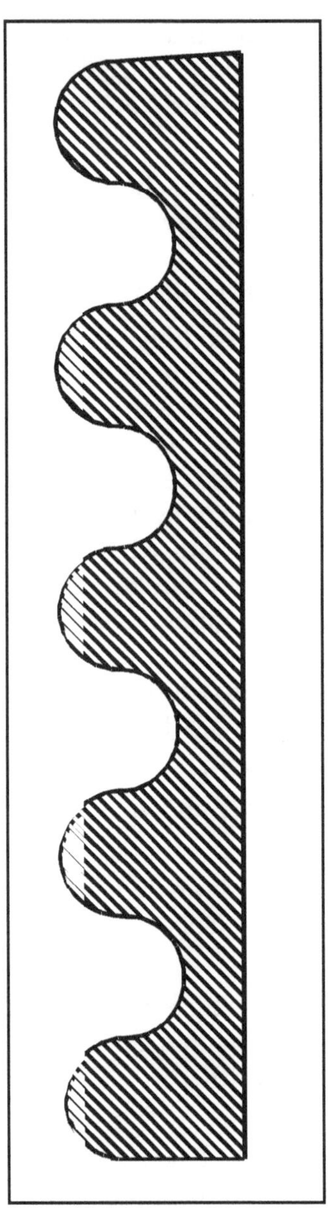

Fig. 1K Knuckle threads

Labeled: Equipment or materials to which has been attached a label, symbol or other identifying mark of an organization acceptable to the "authority having jurisdiction" and concerned with product evaluation, that maintains periodic inspection of production of labeled equipment or materials and by whose labeling the manufacturer indicates compliance with appropriate standards or performance in a specified manner.

Laboratory: A place where scientific tests, experiments, analysis, etc., are carried on.

Ladder: An aid to climbing. Usually made of two parallel uprights connected by regularly spaced rungs.

Ladle: Receptacle used for taking the molten metal from the cupola, in transporting it, and in pouring it into the molds. Ladles are of various shapes and sizes with capacities from 25 pounds to 100 tons.

Lag: A retardation. A sine curve lags another curve when its minimum and maximum points are reached later than the similar points on the other curve.

Laminar flow: Streamline flow of an incompressible, viscous, Newtonian fluid; all particles of the fluid move in distinct and separate lines.

Lancashire boiler: A cylindrical steam boiler consisting of two longitudinal furnace tubes, which have internal grates at the front. It is horizontal smoke-tube boiler. The size ranges from a shell 2 m dia x 6 m long to 3 m dia x 10 m long. The working pressures are up to 20 bars through the general range of 7 to 15 bars. The ratio of heating surface to grate area varies from 24 to 30. Efficiency is about 65 % without economizers and 75 % with economizers.

Land packer (culti packer): It is tandem combination of two Cambridge rollers offset by half a disc or cambridge and cross kill rollers. *See* Fig. 2L.

Landing: 1. The act of terminating flight in which the aircraft is made to descend, lose flying speed, establish contact with the ground and finally come to rest. 2. A platform introduced at some point in a stair run; used to change direction of stair or to break the run.

Lapping: The process of fitting one surface to another by rubbing them together with an abrasive material between the two surfaces.

Laser beam welding: It gives a very concentrated beam of energy. The energy density of laser beam in very high about 80 kW/ mm^2, the beam is narrow that the width of heat-affected zone of the weld is only about 0.05 to 0.1 mm. Used for welding of thin sheet of materials of high speed.

Latent energy: Internal energy associated with the phase of a system.

Latent heat: 1.The amount of energy absorbed or released during a phase-change process. 2. The amount of heat absorbed or evolved by 1 mole, or a unit mass, of a substance during a change of state (such as fusion, sublimation or vaporization) at constant temperature and pressure.

Latent heat of fusion: The amount of energy absorbed during melting and is equivalent to the amount of energy released during freezing.

Latent heat of vaporization: is the amount of energy absorbed during vaporization and is equivalent to the energy released during condensation.

Lateral strain: A strain, which bears against the side of a structure; a transverse strain.

Lateral thrust: The pressure of a load, which extends to the sides.

Lathe: A machine for shaping a work piece by gripping it in a holding device and rotating it under power against a suitable cutting tool for turning, boring, facing, or threading.

Layout: A working plan or diagram of a job.

Lead: 1. An electrical conductor which projects from an electrical device and to which an electrical connection is made. 2. To be in advance. One sine curve leads another when its minimum and maximum points are reached ahead of the same points on the other curve. 3. The distance a screw advances when given a single complete turn. In a single thread the lead is equal to the pitch; in a multiple-thread screw the lead equals the multiple times the pitch; i.e., in a triple thread the lead is three times the pitch. 4. A blue-gray metal, quite soft, ductile, and malleable. Specific gravity 11.34; melting point 327 ^{0}C. Soluble in nitric acid. Usually found as an ore in combination with sulphur, as lead sulphide.

Lead peroxide (PbO_2) :The lead compound from which the positive plates of electric storage batteries are made.

Leaf spring: 1. A beam of cantilever design, firmly anchored at one end and with a large deflection under a load. Also known as flat spring. 2. A spring made up of several flat plates superimposed one upon another, such as used on automobiles, wagons, cars, etc.

Leak testing: Testing by use of a solution, such as a soap solution, to observe a leak under pressure by the formation of bubbles as gas escapes from the leak.

Lean mixture: An air-fuel mixture that has a relatively low proportion of fuel in relation to air. For example, an air-fuel ratio of 16:1 indicates a lean mixture, compared to an approximately normal air fuel ratio of 14.7:1 for gasoline.

Left hand tool: Side tool ground to an angle on the right-hand side; and which, therefore, cuts from left to right.

Lenz's law: The direction of an induced current is always such that the magnetic field belonging to it tends to oppose the change in the strength of the magnetic field, which is setting up the induced e.m.f.

LEV: Low Emission Vehicle.

Leveling instrument: A device consisting of a sighting tube with a spirit level so attached that when the bubble is in the centerline, the line of sight is horizontal.

Lever: A rigid bar turning upon an axis or fulcrum.

Leverage: The mechanical advantage gained through use of a lever.

Lewis bolt: An anchor bolt; a bolt having a jagged and tapered tail. It is used for insertion into masonry, where it is held with lead.

L-head engine: An engine design in which both valves are located on one side of the engine cylinder, which are operated by pushrods actuated by a single camshaft. *See* Fig. 1L

Light duty vehicle: A motor vehicle manufactured primarily for transporting persons or property and having a gross vehicle weight of 6000 lb (2720 kg) or less.

Lighting arrester: A device that causes lightning to pass off to the earth thus serving to protect electrical machines.

Lighting efficacy: The amount of light output in lumens per watt of electricity consumed.

Lilly controller: A device on steam and electric winding engines that protects against over speed, over wind, and other incidents injurious to workers and the engine.

Lime or calcium oxide (CaO)**:** Obtained by the action of heat on limestone. It has many uses in the building arts.

Limestone or calcium carbonate ($CaCO_3$)**:** A very commonly used stone for buildings of the better type; also used for making lime.

Limit control: In machine-tool operation, a sensing device, which terminates motion of the work piece or tool at prescribed points.

Limit governor: A mechanical governor that takes over control from the main governor to shut the machine down when speed reaches predetermined excess above the allowable rate. Also known as tapping governor.

Line of action: The line of action of a force is the direction in which it acts upon a material point, which line must be straight.

Line of pull: It is an imaginary straight line passing from the centre of resistance through the clevis to the centre of pull (power). *See* Fig. 3L.

Linear strain: The ratio of the change in the length of a body to its initial length. Also known as longitudinal strain.

Liner: A replaceable tube to fit inside an engine cylinder, a bushing for a bearing, or the like.

Link: 1. One of the loops of a piece of chain. 2. Component in mechanical linkage

Linseed oil: Oil from the linseed or flaxseed; used principally in paints.

Liquefied hydrogen: Hydrogen cooled to 20.3 K (-423 °F; -253 °C) and ambient pressure becomes a liquid.

Liquefied natural gas (LNG): A motor fuel composed of natural gas that has been liquefied. Liquefied natural gas cooled to 111 K (-259 °F; -162 °C) and ambient pressure becomes a liquid.

Liquefied petroleum gas (LPG): A by-product of natural gas processing or crude oil refining. It consists mainly of propane (over 90 percent), and thus LPG is usually referred to as propane. However, it also contains varying amounts of butane, propylene, and butylenes.

Liquid cooled engine: An internal combustion engine with a jacket cooling system in which liquid, usually water, is circulated to maintain acceptable operating temperatures of machine parts.

Litre (L): Metric measure of capacity corresponding to a volume of 1000 cm^3 (or 1 dm^3). Related units are the millilitre (ml) at 1000 to the litre, and the kilolitre (kl) at 1000l. Engine displacement is often given in cubic centimeters (cc) for smaller engines, and in litres for larger engines.

Live load: A moving or repeated load, which is not constant in its application.

Loam: A mixture of sand and clay used in molding.

Lock ring (wheel) **:** A removable flange or ring used to lock the tyre on to the rim.

Locomotive: A self-propelling machine with flanged wheels, for moving lads on railroad tracks; utilizes fuel (for steam or internal combustion engines), compressed air, or electric energy.

Logarithm: The exponent of the power to which a fixed number, the "base", must be raised to produce a given number.

Loose pulley: The idler or carrier pulley of a pair, on which the belt runs when the machine, which the belt has to drive, is not in use. When the machine has to drive is not in use. When the machine has to be driven, the belt is shifted from the loose to the fast pulley.

Low carbon steels: Those containing less than 0.30 per cent carbon. Such steels can be casehardened but cannot be tempered.

Low gear: The arrangement of change gears which provides the slowest forward speed.

Lower flammable limit (LFL): The LFL of a gas is the lowest gas concentration that will support a self- propagating flame when mixed with air and ignited.

Lower heating value or LHV : The heating value of the fuel when the water in the combustion gases is a vapour. Efficiencies of cars and jet engines are normally based on lower heating values since water normally leaves as a vapour in the exhaust gases, and it is not practical to try to recuperate the heat of vaporization.

Low-pressure cylinder: Those cylinders with a marked service pressure below 900 psi .

Low-tension: The primary circuit (6 volts).

Lubricant: Oil, grease, graphite, and in general, anything of the sort used to overcome friction and to permit a freer action of parts. Also used to exert a cooling action on tool and material that is being cut. The lubricant used for this purpose also removes chips and imparts smoother finish to the parts worked upon.

Lubrication: The act of applying lubricants.

Lubrication system: It is an arrangement of mechanism and devices, which maintains supply of lubrication oil to the rubbing surface of an engine at correct pressure and temperature. Three common systems of lubrication used on stationary engines, tractor engines and automobiles are: 1.Splash system, 2. Forced feed system and 3. Combination of splash and force feed system. See splash system, forced feed system and combination of splash and force feed system

Lug: A projection of irregular shape as from a casting.

Lumber: Timber cut to size in marketable form, as boards, planks, etc.

—□—□—

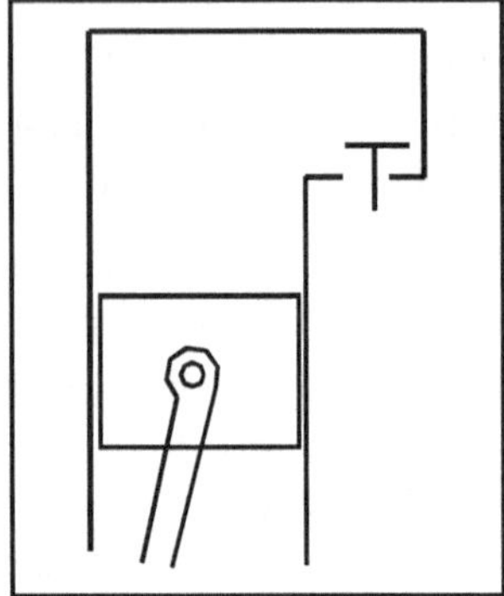

Fig. 1L L- head engine

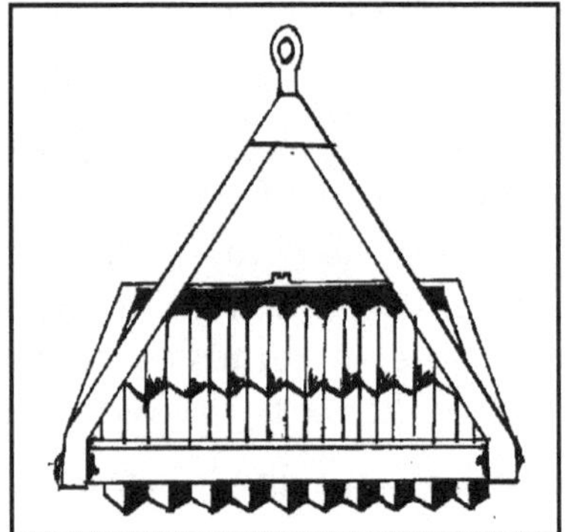

Fig. 2L Land packer (Culti-packer)

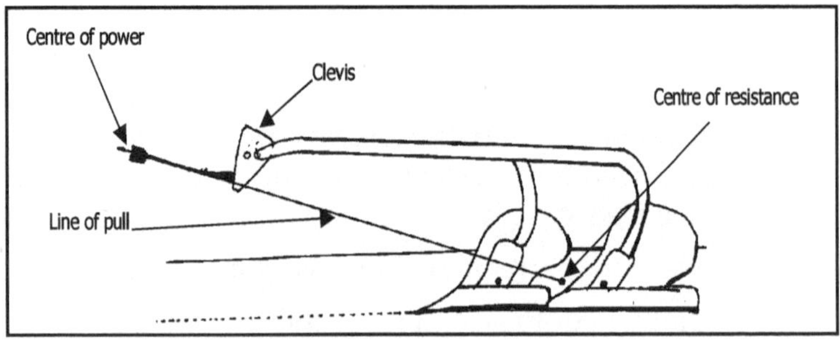

Fig. 3L Line of pull

Mach number: Named after the Austrian physicist Ernst Mach (1838-1916), is the ratio of the actual velocity of the fluid (or an object in still air) to the velocity of sound in the same fluid at the same state.

Machine: 1. A combination of rigid or resistant bodies having definite motions and capable of performing useful work. 2. A device for transforming or transferring energy.

Machine drawing: A mechanical drawing of a machine or machine parts provided with notes and dimensions for workshop information.

Machine drilling: The drilling of work under a power-driven machine.

Machine molding: The use of special machines in the preparation of molds for the production of castings.

Machine rating: The amount of power a machine can deliver without overheating.

Machineable: Material capable of being finished by tools or cutters in or on a machine tool.

Machinery: A group of machines; also, the working parts of an engine or machine.

Machining: Performing various cutting or grinding operations on a piece of work.

Macroscopic: Forms of energy are those a system possesses as a whole with respect to some outside reference frame, such as kinetic and potential energies.

Magazine: 1. A protected building or room for the storage of explosives or ammunition. 2. A paperbound periodical in book form, usually issued monthly. 3. A part of a composing machine in which matrices or letters are stored ready for assembling into lines.

Magnesium: A very light metal with specific gravity of 1.74. It is never used alone but alloyed with aluminum or other metals to produce lightweight airplane parts, etc. The metal ignites very easily and precautions must be taken against fire during machining operations. It is also used in flashlight powders, fireworks, and as a deoxidizer.

Magnet: A body or substance, which has the property of attracting particles of iron to itself. Magnets are of the horseshoe or bar type.

Magnet core: Usually a soft iron center on which the turns of wire are wound to produce an electromagnet.

Magnetic brake: A friction brake under the control of an electromagnet.

Magnetic circuit breaker: An electromagnetic device for opening a circuit.

Magnetic circuit: The path of magnetic lines of force in a magnetic substance or in magnetic or electric apparatus.

Magnetic coil: The winding of an electromagnet. A coil of wire wound in one direction, producing a dense magnetic field capable of attracting iron or steel when carrying a current of electricity.

Magnetic field: The space in the vicinity of a magnet through which magnetic forces act.

Magnetic induction: The number of magnetic lines or the magnetic flux per unit of cross-sectional area perpendicular to the direction of the flux.

Magnetic poles: The north and south poles of any bar or horse shoe magnet. There are the spots on a magnet where the lines of force are heaviest and most numerous and the magnetic force is therefore greatest.

Magnetic switch: A switch actuated by the magnetic pull of a coil and core on a contact disc; used as one kind of starting-motor switch.

Magnetic whirl: The magnetic field in the form of a circular motion about a wire or conductor carrying current electricity.

Magnetism: 1. The property or iron and steel as well as of coils of wire carrying electricity which causes them to attract or repel other units of iron and steel which may or may not be magnetized. An alloy of aluminum, nickel and cobalt, called alnico, can also be magnetized. 2. The science that treats of the conditions and laws of magnetic force.

Magneto: A device consisting essentially of permanent magnets and an armature, for generating electricity by electromagnetic induction.

Magneto ignition system: In this spark ignition system, a high tension magneto is used to generate an electric current for producing spark. The main components of magneto ignition system are permanent magnet, soft iron core, primary and secondary winding, breaker point and condenser. *See* Fig.1M.

Main bearings (automobile) **:** The main bearings in an automobile engine are those, which carry the crankshaft.

Main shaft: It is the line of shafting, which receives its power directly from the engine or motor and transmits power to other parts.

Main shoe: The non-reciprocating part to support the inner end of the cutter bar. It is also known as inner shoe.

Main transmission: A combination of gears and shafts to transmit the power from engine of a tractor to its wheels or tracks.

Maintenance instruction: A document which explains the maintenance and adjustment required, and the manner of under taking work on the equipment at the operator's level.

Maize header effective width: The average distance between centerlines of the adjacent picking units multiplied by the number of units. Where the header width is adjustable. Maximum and minimum dimension shall be stated. Effective width shaft be expressed in millimeters and the number of picking units shall be stated.

Maize planter: A machine of planting of maize seeds in the seedbed.

Malfunction: Improper or incorrect operation.

Mallet: A wooden hammer. *See* Fig. 2M.

Mandrel: A shaft or spindle on which an object may be fixed for rotation; e.g., when a piece is to be turned in a lathe, it may be carried on a "mandrel" which is supported by the lathe centers. The terms mandrel and arbor are often used interchangeably.

Maneuverability: That quality in a machine, which permits ease of handling.

Manganese bronze: An alloy containing 55-60 per cent copper; 38-42 per cent zinc, and small amounts of tin, manganese, aluminum, iron, and lead. Used for parts requiring strength and toughness.

Manhole (boiler)**:** It is an opening big enough for a man to go inside the boiler.

Manifold: 1. A pipe or casting with multiple openings used to connect various cylinders to an inlet or outlet. 2. A device with several inlet or outlet passageways through which a gas or a liquid is gathered or distributed.

Manometer: A device based on the principle that an elevation change Ä z of a fluid corresponds to a pressure change

ÄP/ ñg, which suggests that a fluid column can be used to measure pressure differences. The manometer is commonly used to measure small and moderate pressure differences.

Manual shutoff valve: A hand valve used to isolate a component or circuit.

Manually operated winnowing fan: A winnowing fan operated by manual power either by hand or by pedals.

Manually operated: Operated by human beings.

Manure application: Process of spreading manure on the soil and, in some cases, mixing it with the soil.

Marble: A kind of limestone ranging in color from white to dark gray and to brown; widely used for both interior and exterior finish of buildings; also for switchboard panels.

Margin: 1. A space, along an edge or a bounding line; a border. 2.The spaces on the edge of a printed sheet, which may be blank or contain notes.

Marking gauge: A woodworker's gauge, used for scribing lines parallel to the edge of a board. It consists of a bar with an inserted pin or scribe, and a sliding head, which may be adjusted by means of a thumbscrew.

Marking machine: A machine for stamping trademarks, patent dates, etc., on cutlery, gun barrels, etc. Stamps are usually on rolls and are rolled into the work.

Masonry: The work of one who builds with stone, brick, etc.

Mass airflow sensor (MAF): A sensor that measures the volume of air entering an engine, usually in grams per second (g/sec). Used in onboard computer systems.

Mass flow meter: A device for measuring the mass flow of gases.

Mass flow rate: The amount of mass flowing through a cross section per unit time.

Mass fraction: The ratio of the mass of one component in a mixture to the total mass of the mixture.

Mass of a system: It is equal to the product of its molar mass M and the mole number N.

Mast: The component that provides location of the upper hitch point on the implement.

Mast adjustment: The usable range of pitch of the mast in vertical plane. It is measured as the maximum and minimum heights of lower hitch points above the ground between which a mast can be adjusted to any inclination between the vertical and 100 to the vertical toward the rear.

Mast height: The vertical distance between the upper hitch point and the common axis of the lower hitch points.

Master cylinder: The fluid cylinder carrying the piston connected to the brake pedal (by means of which the brakes are applied through pressure on the foot pedal) in the hydraulic braking system. *Refer* Fig. 6H.

Master gage: A locating device with fixed hole locations or part positions; locates in three dimensions and generally occupies the same space as the part it represents.

Maximum belt tension: The total of the starting and operating tensions in a conveyor.

Maximum drawbar pull: The maximum horizontal drawbar pull, at a drawbar height recommended by the manufacturer, corresponding up to 15 percent wheels slip on standard concrete test track for wheel tractors and up to 7 % track slip for track laying tractors on earthen test track which a tractor is able to sustain in the line of its longitudinal axis.

Maximum permissible driving time: The maximum elapsed time that may be used to complete drying any portion of the grain without undesirable change in grain quality.

Maximum travel speed: The maximum attainable speed of the unladen tractor, where the engine speed governor has been set at the maximum top idle speed rating recommended by the manufacturer.

Maximum voltage: The highest voltage reached in each alternation of an alternating e.m.f.

Mean effective pressure (MEP) : A fictitious pressure, if acted on the piston during the entire power stroke, would produce the same amount of net work as that produced during the actual cycle. The mean effective pressure can be used as a parameter to compare the performances of reciprocating engines of equal size. The engine with a larger value of MEP will deliver more network per cycle and thus will perform better.

Mechanical advantage: The ratio of the force produced by a machine such as a lever or pulley to the force applied to it. Also known as force ratio.

Mechanical agitation (plant protection appliances): Agitation of the spray mixture, dust or granules by means of mechanically operated stirrer inside the tank or hopper.

Mechanical analysis: Mechanical separation of soil, sediment, or rock by sieving, screening, or other means to determine particle-size distribution.

Mechanical brakes: Any brake system where pressure on the brake pedal is transmitted to the brake shoes on each wheel by a combination of rods, levers, cams, or cranks. *Refer* Fig. 10B

Mechanical duster: An appliance used for mechanical dusting.

Mechanical dusting: Distribution of dust by mechanical means

Mechanical efficiency: 1.Ratio between input power and output power. M.E.= output power / input power. 2. In an engine, the ratio between brake horsepower and indicated horsepower.

Mechanical engineer: One who is expert in the design, construction, and use of machinery or mechanical devices.

Mechanical equilibrium: A system is in mechanical equilibrium if there is no change in pressure at any point of the system with time.

Mechanical impedance (soil)**:** The mechanical resistance of the soil to the movement of plant roots or tillage tools.

Mechanical properties: Properties of a material that pertain to its elastic and plastic behavior when force is applied: yield strength, ultimate strength, elongation, hardness, etc.

Mechanical stability: The degree of resistance of the soil to the deformation of breakdown by a specified mechanical force.

Mechanical strength: The mechanical resistance of the soil to deformation or separation by forces.

Mechanical agitator: A device used for mechanical agitation.

Mechanize: 1. To substitute machinery for human or animal labour. 2. To produce or reproduce by machine.

Mechanized: System, which is moved either by engine power, tractor power, waterpower or hand power on wheels or skids. Generally considered as any type of system that can be moved without carrying manually.

Medium volume spray: Spray volume more than 56 and less than 560 liters per hectare.

Mega Pascal (MPa) : The unit of pressure equal to 10^6 Pascal. ($1Pa=1N/m^2$).

Melting line: Separates the solid and liquid regions on the phase diagram.

Melting point: The temperature at which a solid becomes a liquid or vice versa.

Mercury vapour lamp: The type of lamp perfected by Cooper Hewitt, in which light is produced by passing a current through mercury vapour.

Metal arc welding: An arc-welding process wherein the electrode supplies the filler metal in the weld.

Metal: In the trades the term is used not only to designate the elementary metallic substance but those refined or partially refined products of the ores, which have such characteristics as ductility, malleability, fusibility, etc.; also used as a general classification to include a wide variety of alloys.

Meter (m): Basic metric unit of length equal to 3.28 feet, 1.09 yards or 39.37 inches. Related units are the decimeter (dm) at 10 per meter, the centimeter (cm) at 100 per meter, the millimeter (mm) at 1000 per meter and the kilometer (km) at 1000 meters.

Metering feed mechanism: The mechanism of seed drill, planter or fertilizer drill, which delivers seeds or fertilizer from the hopper at selected rates. *Refer* Fig. 10C, Fig. 20C and Fig. 8F.

Metering orifice: A fixed opening alongside the venturi through which gasoline flow is regulated to supply fuel on various demands.

Metering pin: A pin, which seats itself in the metering orifice and regulates the flow of gas through it.

Metering rod; The rod linked to the throttle arm by which the gasoline flow is regulated.

Methanol (CH_3OH): Methyl alcohol is the simplest of the alcohols. It has been used, together with some of the higher alcohols, as a high-octane gasoline component and is a useful automotive fuel in its own right.

Metric SI: It is from Le System International d' Unités, which is also known as the International System, is based on six fundamental dimensions. Their units, adopted in 1954 at the Tenth General Conference of Weights and Measures, are: meter (m) for length, kilogram (kg) for mass, second (s) for time, ampere (A) for electric current, degree Kelvin (K) for temperature, candela (cd) for luminous intensity (amount of light), and mole (mol) for the amount of matter.

Metric system: A decimal system of weights and measures based on multiples of ten. First used in France, now in used universally in scientific work.

Mica: A transparent mineral which readily separates into more or less elastic layers; it is an excellent insulator or non-conductor of electricity.

Micrometer: A measuring instrument for both external and internal measurement in thousandths and sometimes tenths of thousandths of inches or millimeter. *See* Fig.3M.

Microscope: An instrument containing one or more lenses to permit the observations and study of minute objects, which would be invisible without such aid.

Microscopic: Forms of energy those are related to the molecular structure of a system and the degree of the molecular activity, and they are independent of outside reference frames.

Middle breaking: The use of lister in the manner that opens the furrow midway between two previous rows of plants

Miles per hour (mph): Standard measure of speed. Equal to 1.6 km/h.

Millet thresher: Equipment used for threshing of millet crop.

Milligram (mg): Metric unit of weight or mass equal to 1/1000 of a kilogram. The weight of 1 mL or 1 cc of water at atmosphere pressure.

Millimeter (mm): Metric unit of length, equal to 0.04 inch, or 25 mm per inch. There are 1000 millimeters in a meter.

Millimeter of water: A unit of pressure, equal to the pressure exerted by a column of water 1 millimeter high with a density of 1 gram per cubic centimeter under the standard acceleration of gravity; equal to 9.81 m/sec^2. Abbreviated as mm H_2O.

Milling: Mechanical treatment of materials to produce a powder, to change the size or shape of metal powder particles, or to coat one powder mixture with another.

Milliwatt: A unit of power equal to one-thousandth of a watt. Abbreviated as mW.

Minimum clearance diameter (tractor)**:** The diameter of the smallest circle described by the outermost point of the projections of the tractor while executing its sharpest practical turn.

Minimum tillage: The minimum soil manipulation necessary to meet tillage requirements for crop production.

Minimum turning diameter(tractor)**:** The diameter of the circular path described by center of tyre of the outermost wheel of the tractor while executing its sharpest practical turn.

Misfiring: Failure of an explosion to occur in one or more cylinders while the engine is running. It may be a continuous or intermittent failure.

Mist: The distribution of droplets with a VMD value in range of 50 and 100 μm.

Mobile thresher: The thresher, which performs the threshing operation while the machine itself is moving.

Modification: An alteration; a change from the original.

Modulate: Automatically governing the rate of fuel flow by a control, which is temperature sensitive in order to maintain a constant temperature at the location of the sensing device.

Modulus of elasticity: The ratio of the increment of some specified form of stress to the increment of some specified form of strain, such as Young's modulus (ratio of normal stress to longitudinal strain), the bulk modulus (normal stress to volume strain), or the shear modulus/modulus of rigidity (ratio of tangential stress to shear strain). Also known as coefficient of elasticity; elasticity modulus; elastic modulus.

Moisture density: Mass of water per unit volume. When expressed in grams per cubic centimeters, it is numerically equal to the percentage of total space occupied by the water.

Moisture percentage (soil)**:** The percentage of water in soil and plant usually on a dry weight basis. If otherwise expressed, the basis should be stated.

Moisture tension: The equivalent negative gauge pressure or suction in the soil moisture. Soil moisture tension is equal to the equivalent negative or gauge pressure to which water must be subjected in order to be in hydraulic equilibrium through a porous permeable wall or membrane with the water in the soil.

Molar analysis: A way to describe the composition of a mixture that is accomplished by specifying the number of moles of each component.

Molar mass: The mass of one mole (also called a gram-mole, abbreviated gmol) of a substance in grams, or the mass of one kmol (also called a kilogram-mole, abbreviated kgmol) in kilograms. In English units, it is the mass of 1 lbmol in lbm. Notice that the molar mass of a substance has the same numerical value in both unit systems because of the way it is defined.

Mole fraction: The ratio of the number of moles of one component in a mixture to the total moles of the mixture.

Note that for an ideal-gas mixture, the mole fraction, the pressure fraction, and the volume fraction of a component are identical.

Molecular weight: Sum of the weights of atoms making up a molecule.

Molecule: The smallest particle of a substance that retains all the properties of the substance; it is composed of atoms.

Mortar: A heavy-walled vessel in which substances are ground with a pestle. (Masonry) A mixture of sand and slaked lime for joining bricks, stone, etc.

Motive power: The particular source of energy, which is applied to activate a prime mover or a machine.

Motive pressure regulator: A pressure regulator that reduces gas pressure from the high-pressure conditions within the fuel storage system to an intermediate pressure, typically around 175 psig.

Motor: 1. A machine for transforming electrical energy into mechanical energy. 2. An internal combustion engine, as the gasoline engine of an automobile.

Motor efficiency: The ratio of the mechanical energy output of a motor to the electrical energy input. The full-load motor efficiencies range from about 35 percent for small motors to over 96 percent for large high-efficiency motors.

Motor starter: A variable resistance box connected in series with a motor for starting duty. As the speed of the motor increases, the resistance is decreased until it is entirely cut out of circuit.

Motor torque: Turning effort. The turning moment or pull, which causes rotation, or tendency to rotate, in a motor.

Mould board plough: A plough with share and mould board, which cuts furrow slice, lifts and inverts it.

Mould board: The curve part, which lifts and turns the furrow slice. There are different types of mould board as per the requirements as 1. General purpose, 2. Stubble, 3. Sod or breaker and 4. Slat. Refer general purpose, stubble, sod or breaker and slat.

Mounted: Attached to a tractor so as to form a single unit and to be completely carried on the tractor when out of operation.

Moving-coil galvanometer: A sensitive instrument in which a movable coil supported in a strong magnetic field, set up by a permanent magnet, indicates or detects small currents of electricity flowing through the coil.

Mower: A machine to cut herbage crops and leave them in swath. The different types of mowers are as 1. Cylinder mower, 2. Reciprocating mower, 3. Horizontal rotary mower, 4. Gang mower and 5. Flail mower. Refer cylinder mower, reciprocating mower, horizontal rotary mower, gang mower and flail mower.

Mower attachment: A cutting means designed to be easily detached from the vehicle generally to allow vehicle to be used without the mover or with other attachment.

Mph: Miles per hour.

Muffler: A chamber attached to the end of the exhaust pipe, which allows the exhaust gas to expand and cool. It is usually fitted with the baffles or porous plates and serves to subdue much of the noise created by the exhaust.

Muffler silencer: Device for reducing engine exhaust noise and directing the exhaust gases.

Mulch tillage: Preparation of soil in such a way that plant residues or other mulching materials are specially left on or near the surface.

Mulching: Applying a layer of organic matter on the surface of the soil.

Mulching mower: A rotary mower having a fixed or optional configuration without openings in the mower housing above the plane of the blade for the discharge of grass clippings.

Multi plane: A biplane or tri-plane; any airplane with two or more main supporting surface placed one above another.

Multicrop thresher: Threshing equipment used for more then one crop with or without minor adjustments.

Multi-outlet control valve: A device enabling the flow of spray mixture to direct to one or more outlets.

Multiple-plate clutches: Clutch with more than three driving and more than two driven plates or discs.

Multiplier: A known resistance used with a voltmeter or a galvanometer or a galvanometer to increase their range.

Multistage compression with intercooling: A compression process where a gas is compressed in stages and cooled between each stage by passing it through a heat exchanger called an intercooler.

Multistage expansion with reheating: requires the expansion process in a turbine be carried out in stages and reheating the gas between the stages such that the work output of a turbine operating between two pressure levels can be increased.

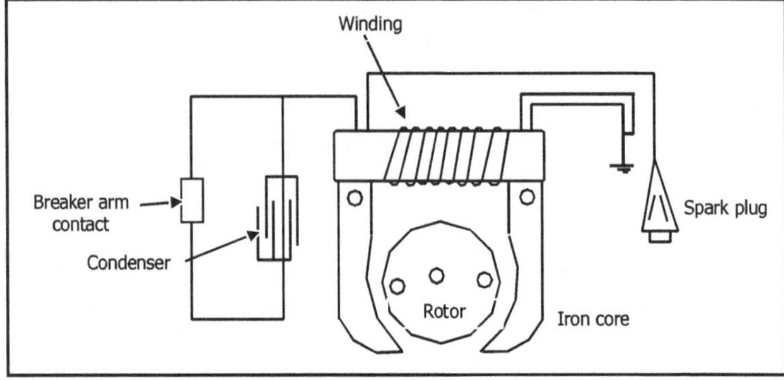

Fig. 1M Magneto ignition system

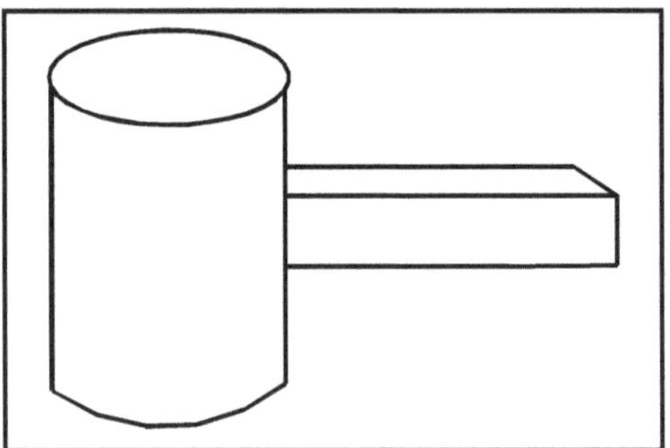

Fig. 2M Mallet

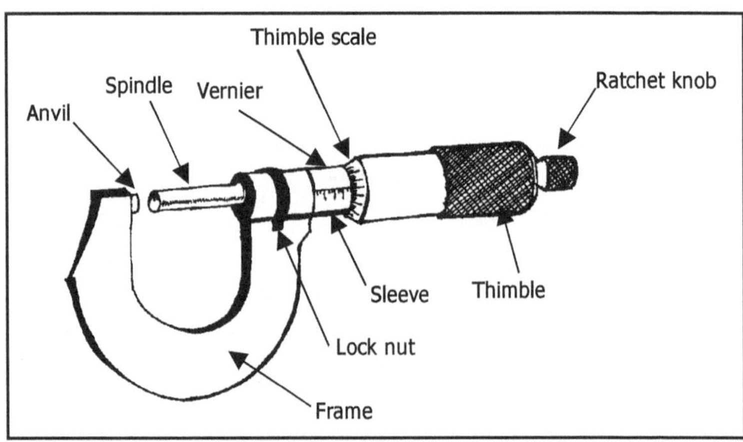

Fig. 3M Micrometer caliper

Natural gas: Produced from gas wells or oil wells rich in natural gas. It is composed mainly of methane, but it also contains small amounts of ethane, propane, hydrogen, helium, carbon dioxide, nitrogen, hydrogen sulphate, and water vapour. It is stored either in the gas phase at pressures of 150 to 250 atm as CNG (compressed natural gas) or in the liquid phase at 162° C as LNG (liquefied natural gas).

Natural-draft cooling tower: Uses the naturally occurring density gradients between the inside air-water vapour mixture and the outside air which create an airflow from the bottom to the top of a wet cooling tower.

Needle valve: A valve in which the flow of liquid or gas is regulated by the adjustment of a pin or needlepoint, which sets in a cone-shaped depression having a small hole at the bottom.

Negative: One of the two terminals in an electrical circuit, presumably the one through which the current returns to the source.

Negative camber: The condition existing when two tyres are closer together at the top than at the bottom.

Negative charge: An excess of electrons over protons. For example, the negative terminal of a cell has more electrons (negative charges) than the positive terminals, therefore, it is negatively charged.

Negative plate: Any one of the plates used in the negative groups of the storage-battery element; presumed to be the plates to which the electrical circuit flows from the external circuit.

Neon light: A type of lamp with two electrodes instead of a filament. The ionization of neon gas contained in the tube produces illumination. Such lights are extensively used in advertising displays.

Neon-light ignition timing: The use of a synchroscope for determining the firing position of the ignition timing when the engine is running.

Net traction: Force in direction of travel developed by the traction device and transferred to the vehicle.

Neutral: the gear-shift position in which engine power is not transmitted to the driving axle, i.e., the speed change gears are disengaged. Neither positive nor negative.

Neutral axis: In a simple beam, the top fibers are always in compression and the lower fibers always in tension. There must, therefore, be some point at which the fiber is called the "neutral axis" of the section.

Neutral position: That position of the gear-shift lever, which places the change gears out of contact permitting the engine to run idle without giving motion to the car.

Newton (N): Metric unit of force that, acting on a mass of one kilogram, increases its velocity by one meter per second every second along the direction that it acts. Named for English mathematician Sir Isaac Newton (1642:1727) who had the greatest single influence on theoretical physics until Albert Einstein. He discovered the binomial theorem and differential calculus. In his major treatise *Principia Mathematica* (1687), generally considered the greatest scientific work ever written, he gave a mathematical description of the laws of mechanics and gravitation, and applied these to planetary and lunar motion. He also discovered that white light is made up of a mixture of colors, built the first reflecting telescope and revolutionized the minting of coins to prevent counterfeiting.

Newton's laws of motion: 1st law - Every body continues in a state of rest or in uniform motion in a straight line, except as it is compelled by a force to change its state of rest or motion. 2nd law - If a body is acted upon by several forces, it is acted upon by each of these as if the others did not exist. 3rd law – To every action there is always an equal reaction, i.e., if a forces acts to change the state of motion of a body, the body offers a resistance equal and directly opposite to the force.

Nichrome: An alloy of nickel, chromium and iron (Ni-65%, Cr-15% and Fe-20%). It has a high resistance against oxidation and corrosion and widely used for electrical resistance wires for electrical heating devices like heating elements.

Nickel copper: A nickel and copper alloy used in making acid-resistant casting and bearing bronzes.

Nickel plating: The depositing of a coating of nickel on metallic surface. Accomplished in nickel salt bath through which an electric current of low voltage is passed.

Nickel steel: A steel containing about 3 percent nickel. It is very strong and tough when properly heat-treated.

Nitriding: It is the process of forming an extremely hard skin up to about 0.25 mm deep on steel leaving a soft tough core by heating in ammonia gas which gives up nitrogen to react upon the aluminum included in the special Nitriding steel.

Nitrogen (N_2): A colorless, tasteless, odorless gas that constitutes 78% of the atmosphere by volume and is a part of all living tissues.

Nitrogen oxides (NO_x): Any chemical compound of nitrogen and oxygen. Nitrogen oxides result from high temperature and pressure in the combustion chambers of automobile engines and other power plants during the combustion process. When combined with hydrocarbons in the presence of sunlight, nitrogen oxides form smog. A basic air pollutant; automotive exhaust emission levels of nitrogen oxides are regulated by law.

No tillage planting: Direct planting in essentially unprepared seedbeds.

Noble metals: Metals (such as gold, silver, platinum, and palladium) that do not oxidize readily or enter into other chemical reactions. These metals will promote reactions between other substances. Platinum and palladium are used as the catalysts in catalytic converters.

Nodes: The points of constant potential located between loops of vibration in a circuit through which and oscillatory current is passing.

Nomenclature: A list of names of terms common to any particular art or science.

Nominal spray angle: Spray angle obtained at a reference pressure which characteristics a given type of nozzle.

Non-combustible material: A material that in the form in which it is used and under the conditions anticipated, will not ignite, burn, support combustion, or release flammable vapors when subjected to fire or heat.

Non-reacting gas mixture: A mixture of gases not undergoing a chemical reaction and can be treated as a pure substance since it is usually a homogeneous mixture of different gases.

Non-return valve: An automatic device, which permits the flow of a fluid in one direction only.

Non-separable spark plug: A spark plug, which cannot be disassembled.

Normal ploughing: Ploughing up to depth of 15 cm.

Normalizing: The heating of steel above the upper critical temperature and then cooling in air. The purpose of normalizing is to sleeve internal stresses. It aims at a refinement of steel structure to improve the properties of mechanical strength.

Nose: The business end of tools or things. The threaded end of a lathe or milling machine spindle, or the end of a hog nose drill or similar tool.

Nose guard: Device, covering the nose area of the guide bar, for reducing the incidence or severity of the kickback.

Nose radius: The radius measured in the back rake or top rake plane of a cutting tool.

Notched-wheel tuber planter: A manually fed tuber planter, the planting element of which consists of a vertical spacing wheel with self-emptying indentation of cells.

Nozzle: A device that increases the velocity of a fluid at the expense of decreasing pressure.

Nozzle bar: Rigid or flexible tube fitted to the end of nozzles supplying the spray liquid.

Nozzle boss: The leg on the spray boom or spray lance to which the nozzle body or nozzle cap is fitted.

Nozzle efficiency: the efficiency with which a nozzle converts potential energy into kinetic energy, commonly expressed as the ratio of the actual change in kinetic energy to the ideal change at the given pressure ratio.

Nozzle spacing: The distance between center-to-center of adjacent nozzles on a spray boom.

Nuclear energy: is the tremendous amount of energy associated with the strong bonds within the nucleus of the atom itself.

Nuclear energy: The energy released in nuclear reaction.

Nuclear power plant: A power plant in which nuclear energy is converted into heat for use in producing steam for turbines, which in turn drive generators that produce electric power. Also known as atomic power plant.

Nucleus: The positively charged particle that is the center of an atom.

Number drills: Small drills numbered from 1 to 80, with diameters expressed in thousands of an inch. No.1 is 0.228 in. diameter; No. 80 is 0.0135 in. diameter.

Numbering machine: A machine or device for stamping consecutive numbers or checks, tickets, etc.

Numerical control (NC) **:** A system of controlling a machine or process by instructions in the form of number and letters.

Nut: A small block of metal or other material commonly square or hexagonal in shape, having internal threads to receive a bolt.

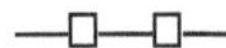

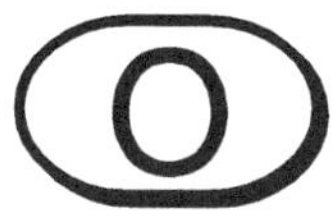

Oak: A hard, durable, and very strong wood used for many purposes like furniture, flooring, decoration etc.

Octane number: It is the percetage of iso-octane (C_8H_{18}) in the mixture of iso-octane and normal heptane (C_7H_{16}), when it produces the same knocking effect as the fuel under test. It is used to indicate the octane rating of a gasoline, describes the anti-knock properties of a fuel when used in an internal combustion engine. Knock is a secondary detonation that occurs after fuel ignition due to heat buildup in some other part of the combustion chamber. When the local temperature exceeds the auto ignition temperature, knock occurs.

Octane rating: A measure of the antiknock properties or engine knock resistance of a gasoline . The higher the octane rating, the more resistant the gasoline is to abnormal combustion.

Octane requirement: The fuel octane number needed for efficient operation (without knocking or spark retardation) of an internal combustion engine.

Odorization: A process of adding a distinctive odor to natural gas so that its presence can be easily detected.

OEM: Original Equipment Manufacturer

Off road vehicle: A conveyance designed to travel on unpaved roads, trails, beaches, or rough terrain and has height and width that may exceed highway legal limits.

Offset: A recess or sunken panel in a wall. A smudge resulting from carrying too much ink on a sheet. Bent out of line in order to clear another structural member.

Offset disc harrow: A disc harrow with two gangs in tandem, capable of being offset to either side of centerline of pull. *Refer* Fig.9D.

Offset plough: A plough with offset beam to work borders and headland.

Ohm: A unit of electric resistance equal to the resistance of a circuit in which a potential difference of one volt produces a current of one ampere.

Ohm's law: States that the current flowing in a circuit (I) is proportional to the e.m.f.(R) and inversely proportional to the resistance or opposition (R). More frequently it is expressed in mathematical from; V = I x R. This law explains the relationship of voltage, current, and resistance in a circuit. One volt of electrical pressure is needed to push one ampere of electrical current through one ohm of resistance.

Ohmmeter: A type of galvanometer, used to measure the amount of resistance (number of ohms) in an electric conductor or circuit.

Oil: A liquid lubricant; made from crude oil and used to provide lubrication between moving parts. In a diesel engine, oil is used for fuel.

Oil bath type air cleaner: It has oil bath and filter element made from wire. The foreign materials in the air get trapped in the oil as air impinges on the oil surface during suction stroke of an engine. Also remaining dust particles are arrested in the wire element as air passes through it before reaching an engine manifold.

Oil control ring (oil ring)**:** 1. A type of piston ring designed to scrape oil from the cylinder wall, that passes through slots in the ring and small holes in the piston wall, and then drains to the crankcase. Such rings are usually placed next above the piston pin bosses. 2. A type of sealing ring, made of a special rubber compound. The Oil ring is placed into a groove to provide the sealing action.

Oil cooler: A small radiator used to cool the oil that lubricates an automotive engine.

Oil furnace: A combustion chamber in which oil is the heat producing fuel.

Oil line: The tubing and devices connected therewith which comprise the circuit for lubricating oil.

Oil pump : Oil pumps are of the gear, vane, or plunger type. They are usually an integral part of the engine. Their purpose is to lift oil from the sump to the upper level in the splash and circulating system, and in forced-feed lubrication they pump the oil to the tubes leading to the bearings and other parts.

Oldham coupling: Mechanism for connecting two parallel shafts whose axes are slightly out of line.

Olefin: An unsaturated hydrocarbon containing one or more double bonds.

Olpad thresher: Animal operated thresher used mainly for wheat consisting of a series of notched discs placed on three axles fixed on wooden or iron frame on which seat and platform are provided. Animal operates this thresher. *See* Fig.2O.

One-way plough: A plough, which turns the furrow slice in the same direction with respect to the forward travel.

On-off Switch: Device for connecting and disconnecting the electric current and thus allowing the motor to started or stopped.

Open circuit: 1. A circuit, which is not electrically completed and in which there is no current. 2. A circuit, which will not function because of the fact that there is a loose connection, a broken wire or similar fault, which prevents current flowing through the circuit.

Open cycle engine: Ordinary gas-turbine engine in which air is drawn in, heated by the combustion of fuel, and discharged together with the combustion products.

Open system or control volume: Any arbitrary region in space through which mass and energy can pass across the boundary.

Operational Mass of Power Tiller: The mass of power tiller in normal working condition with fuel tank and radiator full and lubricants filled to the specified levels.

Operational mass of the tractor: The mass of the tractor in normal working condition with fuel tank and radiator full, lubricants, etc, filled to the specified levels and mass of the driver. (Assumed to be 75kg.)

Operator: A person who actually operates, manipulates or controls the working of a machine .

Operator's handbook: A document, which explains the procedures for the proper operation and maintenance of the equipment.

Optimum tillage: A system of tillage contributing to maximum net return for a crop under given field conditions.

Orchard ladder: Ladder used for facilitating picking of fruits, pruning spraying and other common operation carried out in the gardens.

Oriented tillage: Tillage operations, which are, oriented in specific paths or directions with respect to the sun, prevailing winds, previous tillage action or field base lines.

Orifice: A small opening, or hole, that controls the flow rate of a gas or liquid into a cavity.

Orsat gas analyzer: A commonly used device to analyze the composition of combustion gases. The amounts of carbon dioxide, carbon monoxide, and oxygen are measured on a percent by volume and are based on a dry analysis.

Orthographic (graphics)**:** Relating to the arrangement of views in a mechanical drawing.

Orthographic projection: A system of graphically presenting an object by means of several views, each view showing a face of the object, such as the front view, top view, right-side view, etc.

Oscillating conveyor: A conveyor on which pulverized solids are moved by a pan or trough bed attached to a vibrator or oscillating mechanism. Also known as vibrating conveyor.

Oscillating granulator: Solids size-reducer in which particles are broken by a set of oscillating bars arranged in cylindrical form over a screen of suitable mesh.

Oscillating screen: Solids separator in which the shifting screen oscillates at 300 to 400 revolutions per minute in a plane parallel to the screen.

Oscillating spout fertilizer distributor: A centrifugal fertilizer distributor in which the broadcasting mechanism consists of at least one oscillating spout.

Oscillation: A backward and forward motion as in a pendulum; a vibration.

Oscillograph: An instrument used for studying waveforms of current or voltage either by means of a screen or photograph.

Osmotic pressure: The pressure difference across a semi permcable membrane that separates fresh water from the saline water under equilibrium conditions.

Osmotic rise: The vertical distance saline water would rise when separated from the fresh water by a membrane that is permeable to water molecules alone at equilibrium.

Otto cycle (four stroke cycle): The ideal cycle for spark-ignition reciprocating engines. It is named after Nikolaus A. Otto, who built a successful four-stroke engine in 1876 in Germany using the cycle proposed by Frenchman Beau de Rochas in 1862. The ideal Otto cycle, which closely resembles the actual operating conditions, utilizes the air-standard assumptions. It consists of four internally reversible processes:1-2 Isentropic compression; 2-3 Constant volume heat addition; 3-4 Isentropic expansion; 4-1 Constant volume heat rejection. The four events in the cycle occur in one revolution of crankshaft. *See* Fig. 1O.

Otto engine: An internal combustion engine that operates on the Otto cycle, where the phases of suction, compression, combustion, expansion, and exhaust occur sequentially in a four-stroke-cycle or two-cycle reciprocating mechanism.

Outboard bendix drive: The bendix drive in which the pinion gear moves outward (away from the motor) when the starting switch is closed. This design usually requires an outboard bearing.

Outer shoe: The non-reciprocating pointed attachment at the outer end of the cutter bar caring wheel or skid and grass board

Outer shoe skid: A steel skid fitted underneath an outer shoe to adjust the height of the cutter bar.

Outer shoe wheel: A wheel fitted to an outer shoe to adjust the height of the cutter bar.

Output: Amount of energy delivered to an external device from the source of generation. Work accomplished which is equal to input less various losses.

Output traction power: Power output at traction device calculated from net traction force and velocity in the direction of motion.

Over speed governor: A governor that stops the prime mover when speed is excessive.

Over steer: The tendency of an automotive vehicle to steer into a turn to sharper degree than was intended by the driver; sometimes causes the vehicle's rear end to swing out.

Overall efficiency (power plant)**:** For a power plant is the ratio of the net electrical power output to the rate of fuel energy input. Overall efficiencies are about 25 to 28 percent for gasoline automotive engines, 34 to 38 percent for diesel engines, and 40 to 60 percent for large power plants.

Overflow pipe: A pipe through which excess fluid from a pump is by-passed by the action of a relief valve or pressure regulator.

Overhaul: To take apart, inspect, repair, and reassemble, as a piece of machinery.

Overhead camshaft: A camshaft mounted above the cylinder head.

Overhead spraying: The spraying performed in the field on top of the crop.

Overhead valve: *See* Fig.3O.

Overheat: 1.To become excessively hot. 2. To operate above the manufacturer's recommended temperature range.

Overheating: The heating of an engine beyond the point of safe operation. Usually due to a defective cooling system or poor lubrication.

Overload (electrical)**:** More than a normal amount of current flowing through an electrical device.

Over-running clutch: A clutching device which permits the starting motor armature to drive the engine fly-wheel but which prevents the engine flywheel from driving the starting motor armature, since the clutch over-runs the shaft under that condition.

Oxidation: Burning or combusting; the combining of material with oxygen. Rusting is slow oxidation, and combustion is rapid oxidation.

Oxy acetylene welding: Method of welding in which an oxy-acetylene flame is used to melt the edges of the parts to be joined. In most cases additional metal is fed into the joint from a filler rod of suitable composition.

Oxygen (O_2): A colorless, tasteless, odorless, gaseous element that makes up about 21% of air. Oxygen is capable of combining rapidly with all elements (except inert gases) in the oxidation process called burning (combustion). Oxygen combines very slowly with many metals in the oxidizing process called rusting.

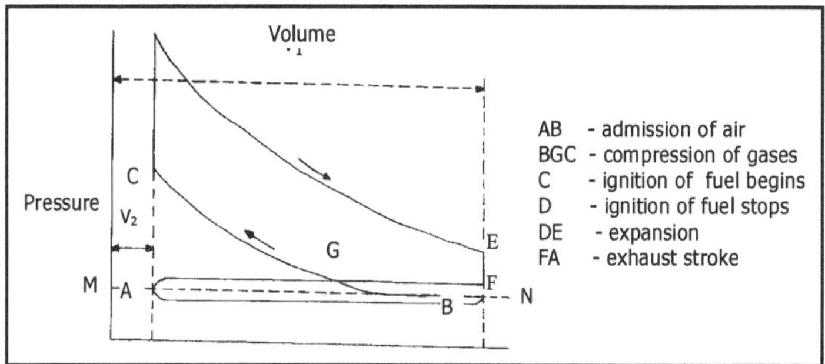

Fig. 1O Otto cycle

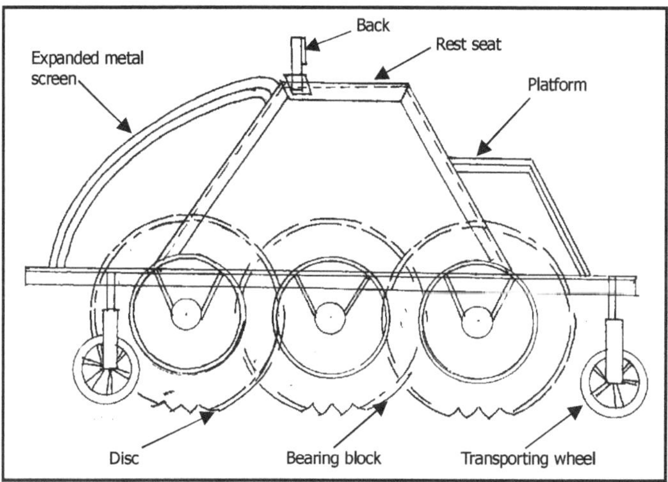

Fig. 2O Olpad thresher

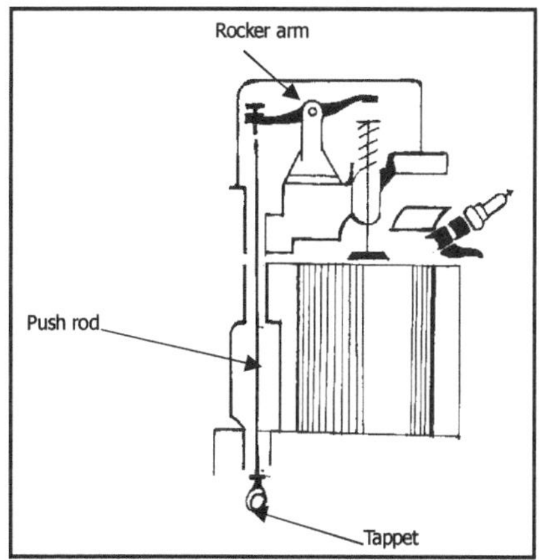

Fig. 3O Overhead valve

Packing: 1. The material, which is enclosed in a stuffing box for the purpose of preventing leakage around a piston rod. 2. The act or process of inserting packing material. 3. Blocking up.

Paddy dehusker: An equipment to remove husk from paddy grain. It is also known as paddy Sheller.

Paddy separator: A machine, which is used for the separation of brown rice from the paddy brown rice mixture.

Paddy thresher, pedal operated: Equipment used for threshing of paddy crop, consisting of mainly a balanced cylinder with series of threshing teeth fixed on slats and worked by a pedal and gear drive mechanism. *See* Fig.1P.

Paddy weeder: Equipment used to uproot the weeds in paddy crop planted in line. *See* Fig.2P.

Pantograph: An instrument for measuring angles, elevations, etc.

Parachute: An umbrella-like device used to retard the descent of a falling body by offering resistance to its motion through the air.

Parallel connection (electrical)**:** The method whereby a number of electrical devices may be connected in to the circuit so that one device may operate independently of another and all at equal voltages or pressures. Also, a method of connecting cells (negative to negative and positive to positive), giving grater amperage than but the same voltage as one cell.

Parallel hybrid: A type of hybrid electric vehicle in which the alternative power unit is capable of producing motive force and is mechanically linked to the drive train.

Parallelogram forces: The result of two forces operating at an angle, which can be enclosed in a parallelogram. The diagonal is the resultant.

Parboiling: A partial cooking of paddy by imparting heat treatment to the soaked paddy by steaming.

Parking brake (vehicle)**:** In an automotive vehicle, a brake that functions independently of the service break and is set after the vehicle has been brought to a stop.

Parking brake (trailer)**:** Assembly of parts by means of which the trailer can be maintained at rest even on slope and in the absence of the driver.

Parking lamp: The term was formerly applied only to those low powered lamps mounted where they could be seen from either direction, showing white from the front and red from the rear. The term now covers both cowl lights and rear lamp taken collectively.

Partial derivative: The change in a function that depends on two (or more) variables, such as $z = z(x, y)$, when allowing one variable to change while holding the others constant and observing the change in the function as another variable is held constant. The variation of $z(x, y)$ with x when y is held constant is called the partial derivative of z with respect to x.

Partial pressure: 1. The product of the mole fraction and the mixture pressure. The partial pressure is identical to the component pressure for ideal gas mixtures. 2. The amount of pressure contributed by an individual gas in a mixture of ideal gases toward the overall, total pressure.

Partial volume: 1. The product of the mole fraction and the mixture volume. It is identical to the component volume for ideal gas mixtures.

Particle: A very small piece of metal, dirt or other impurity that may be contained in the air, fuel, or lubricating oil used in an engine.

Particulates (particulate matter, pm) : Small particles (generally less than 30 microns) of carbon and other pollutants occurring as solid matter in the exhaust systems of vehicles. These particles generally remain suspended in the atmosphere, causing a major pollution problem in many cities. Current EPA regulations address only "respirable" or 'fine" particles that are less than 10 microns in diameter

Parting tool: A narrow blade turning tool used for cutting grooves, recesses and for cutting off.

Partition plate: The horizontal member at the top of a partition wall, serving as a cap for the studs and as a support for joists, rafters, or studs.

Parts catalogue: A document which shows parts which are required to service the equipment. (It is recommended that this should be a separate publication).

Pascal (pa): Metric unit of pressure. Related units are the kilopascal (kpa) at 1000 Pa, the bar at 100,000 Pa and the mega Pascal at 1,000,000 Pa. The Pascal is named for Blaise Pascal, French mathematician and physicist (1623:1662) who discovered that air has weight, confirmed that a vacuum could exist and derived the principle that the pressure of fluid at rest is transmitted equally in all directions. He also founded the theory of probability, and developed a forerunner of integral calculus. $1Pa=1N/m^2$.

Passivated element: A substance that does not readily enter into chemical combination.

Passive solar system: A solar heating or cooling system that operates by using gravity, heat flows, or evaporation rather then mechanical devices to collect and transfer energy.

Patela**:** A wooden plank used for smoothening the soil and crushing the clods. In certain cases weeds can be collected with the help of curved spikes attached behind the wooden board.

Path of a process: The series of states through which a system passes during a process.

Pattern: A model or specimen; something made for copying, or for reproducing similar articles. The model from which metal castings are reproduced.

Pattern making: The making of a model or plan for foundry projects.

Pawl: 1. A hinged or pivoted arm 2. A locking device; a piece of metal fitted into a notch, slot, groove or similar position to hold some other part, having a pointed edge or hook made to engage with ratchet teeth. 3 . The driving link or holding link of a ratchet mechanism, per unit's motion in one direction only.

Payload: 1. A part of the useful load from which revenue is derived, viz., passengers and freight. 2. The uniformly distributed maximum safe load, which can be transported by the trailer.

Peak load: The heaviest load which a generator or system is called on to supply at regular intervals, as once every twenty-four hours.

Peaucellier linkage: A mechanical linkage to convert circular motion exactly into straight-line motion.

Pebble mill: A solids size-reduction device with a cylindrical or conical shell rotating on a horizontal axis, and with a grinding medium such as balls of flint, steel, or porcelain.

Pedal pressure: The force required on the pedal to make it fulfill its function to declutch or set the brake for instance.

Pedestal design: A robot design centered on the vertical axis of a central pedestal, in which the motion of any workpiece is confined to a spherical working envelop.

Pedestrian controlled mower: A grass-cutting machine, either pushed or self propelled, normally controlled by the operator walking behind the unit.

Pedometer: An instrument for recording distance walked.

Peds: Agglomerations of primary soil particles produced by natural processes.

Peeling: The removal of the outer layer of a fruit or vegetable is generally referred as peeling, paring or skinning.

Peen: The small end of the head of a hammer, as the ball peen hammer for metal workers. *Refer* Fig. 3D.

Pcg drum: A cylinder having rows of spike or pegs.

Peltier effect: The cooling effect that occurs when a small current passes through the junction of two dissimilar wires. This effect forms the basis for thermoelectric refrigeration and is named in honor of Jean Charles Athanase Peltier, who discovered this phenomenon in 1834.

Pencil: The high-tension brush and conductor in the high-tension magneto.

Penetrometer: An instrument used to measure soil strength.

Percent deficiency of air: The deficiency of air expressed as a percent of stoichiometric air. For example, 90 percent theoretical air is equivalent to 10 percent deficiency of air.

Percent excess air or **percent theoretical air:** The amount of excess air usually expressed in terms of the stoichiometric air. For example, 50 percent excess air is equivalent to 150 percent theoretical air.

Peristaltic pump: A volumetric pump in which the flow of liquid is achieved by the continuous progression of a flexible tube.

Permanent load: A load which is constant and unvarying; a dead load as the weight of the structure itself or a load imposed, or both taken together.

Permanent magnet: A piece of magnet steel, which retains the acquired property of attracting other pieces of magnetic material after being under the influence of a magnetic field.

Permeability: The arrangement of any determinate number of things or letters in all possible orders, one after another.

Perpendicular: Meeting a given line or surface at right angles.

Perpetual-motion machine (PMM)**:** Any device that violates either the first or second law of thermodynamics. PMM1 is a device that violates the first law of thermodynamics (by creating energy). PMM2 is a device that violates the second law of thermodynamics.

Petrol: British name for gasoline.

Petroleum: A natural oil taken from the earth from which diesel fuel, gasoline, kerosene, benzene, etc. are refined. Used extensively for heating and as a motor fuel oil.

Phase: The time instant when the maximum, zero, or other relative value is attained by an electric way.

Phase angle (electric) **:** The angle, which expresses the phase relation in an alternating-current circuit.

Phase diagram (thermodynamics)**:** The P-T diagram of a pure substance and shows all three phases separated from each other by the sublimation line, vaporization line, and melting line.

Phase equilibrium (thermodynamics)**:** The condition that the two phases of a pure substance are in equilibrium when each phase has the same value of specific Gibbs function. Also, at the triple point (the state at which all three phases coexist in equilibrium), the specific Gibbs function of each one of the three phases is equal.

Phase meter (electric): A meter, which indicates the frequency of the circuit to which it is attached. A frequency meter.

Photoelectric cell: A device, which changes light energy to electrical energy. Also called 'electric eye'. Principal use is for automatic control through light source.

Photostate: A photographic process for rapid duplication of maps, drawing, charts, and records on sensitized paper.

Physical properties: Characteristics that pertain to the nature and composition of a material or object.

Picker wheel mechanism: A feed mechanism for larger seeds like potato in which a vertical plate is provided with radically projected pick - up arms which lifts the seeds by means of jaws or pins and drops them by knock out mechanism.

Pick-up attachment (reaper): A device for gathering the crop from a windrow.

Pick-up attachment width (reaper): The minimum distance including the width of the outermost conveying elements but not including the gather of the flared side sheets expressed in millimeters.

Pick-up baler: A machine to collect the crop or straw from swaths or windrows, to compress it to rectangular shaped mass and to tie firmly with wire or twine for easy handling and storage.

Piezoelectric (or press-electric) **effect:** The emergence of an electric potential in a crystalline substance when subjected to mechanical pressure. This phenomenon, first discovered by brothers Pierre and Jacques Curie in 1880, forms the basis for the widely used strain-gage pressure transducers.

Pigeonhole: Small open compartment, as in the upper portion of a roll-top disk, used as receptacle for letters, envelopes, etc.

Piling: 1. Depositing trees or parts of the trees in orderly piles. 2. Large timbers or poles driven into the ground or the bed of a stream to make a firm foundation.

Pilot bearing: A small bearing in the center of the flywheel, which carries the forward end of the clutch shaft.

Pilot light: A small bulb placed in an electrical circuit so arranged as to light or to show whether that particular circuit is in operation.

Pilot tube: An instrument that measures the stagnation pressure of a flowing fluid, consisting of an open tube pointing into the fluid and connected to a pressure-indicating device. Also called impact tube.

Pin: A small peg or wooden nail. One of various sized fire clay supports for objects fired in a kiln.

Pin punch: A long, slender punch used for driving out tight-fitting pin.

Pin spanner: Used on round nuts having holes in the periphery to receive the spanner pin.

Pincers: A joined instrument with two handles and a pair of grasping jaws for holding an object.

Pinion gear: Usually the smaller gear used to drive a larger gear. May be of any type, for instance, spur, helical or bevel.

Pinion gear and shaft: The pinion gear and shaft, made as unit from the propeller shaft to the rear axle. The unit is supported by two-tapered roller bearing; all are contained in a housing attached to the front of the rear-axle housing.

Pinion shaft carrier: That part of the differential carrier which projects forward from the rear axel and is used as the mounting for the rear axel pinion gear shaft.

Pinwrench: Used on round nuts, which have two holes in their face to receive the pins of the wrench.

Pipe cutter: A tool for cutting wrought iron or steel pipes. The curved end, which partly encircles the pipe carries one or more cutting disks. A screw regulates feed of the cutter as the tool is rotated around the pipe.

Piston: 1. A cylindrical part closed at one end, which is connected to the crankshaft by the connecting rod. The force of the expansion in the cylinder is exerted against the closed end of the piston, causing the connecting rod to move the crankshaft. 2. The plunger, which moves within the cylinder of an engine or pump. The efficiency of compression depends very largely on the proper fitting of the piston. *Refer* Fig. 2D

Piston collapse: A condition describing a collapse or a reduction in diameter of the piston skirt due to heat or stress.

Piston displacement: The volume, which a piston in a cylinder displaces in a single stroke, equal to the distance the piston, travels times the internal cross section of the cylinder.

Piston drill: A heavy percussion type rock drill mounted either on a horizontal bar or on a short horizontal arm fastened to a vertical column; drills holes to 6 inches. (15 centimeters) in diameter. Also known as Reciprocating drill.

Piston engine: A type of engine characterized by reciprocating motion of pistons in a cylinder. Also known as displacement engine; reciprocation engine.

Piston head: The top, closed end of a piston. *Refer* Fig. 2D

Piston lands: The parts of a piston between the piston rings.

Piston pin (wrist pin) : 1. A hollow steel shaft, hardened and ground, which connects the upper end of the connecting rod to the piston.

Piston pump: A volumetric pump in which motion and pressure are applied to the fluid by a reciprocating piston in a cylinder. Also known as Reciprocating pump.

Piston ring: An expanding ring placed in the grooves of the piston to seal off the passage of fluid or gas past the piston. There are two types: compression rings for sealing the compression in the combustion chamber, and oil rings to scrape excess oil off the cylinder wall.

Piston ring cap: The clearance between the ends of the piston ring.

Piston ring expander: A spring placed between the piston ring in a groove to increase the pressure of the ring against the cylinder wall.

Piston ring groove: The channel or slots in the piston in which the piston rings are placed.

Piston rod: The rod, which moves the piston and is connected to the crosshead or crankshaft.

Piston skirt: That part of a piston below the piston pin or rings.

Piston speed: The total distance a piston travels in a given time; usually expressed in feet per minute, metre per minute etc.

Piston stroke: The complete length of travel of a piston in its cylinder. *Refer* Fig. 2D.

Piston valve: A piston operating in a cylindrical case having ports, which are opened and closed by motion of the piston.

Pitch: A tar like product of crude oil distillation; used to seal the cell cover to light to show whether that particular circuit is in operation.

Pitch circle: In toothed gears, an imaginary circle concentric with the gear axis which is defined at the thickest, point on the teeth and along which the tooth pitch is measured. *Refer* Fig.9G.

Pitch ratio: Ratio of propeller pitch (geometrical, unless otherwise stated) to diameter P/D.

Pitman: A rod or arm, which connects a rotary with a reciprocating part; a connecting rod. *See* Fig.3P.

Pitman arm: The steering arm, carrying a ball, attached to the cross shaft of the steering gear. In general, an arm or rod which connects the rotating with a reciprocating part.

Pivot pin: The king pin on which, the front axle steering knuckle body turns.

Plain live axle shaft: An axle shaft, supported in bearings at either end, which carries the driving load on its inner end and the vehicle load on its outer end.

Plan: A draft or form drawn on a plane surface as a top view of a horizontal section.

Planetary gears: A central sun gear (generally mounted on a shaft) surrounded by a set (usually three) of planet gears, which are mounted on a spider and operate within a ring or internal gear.

Plank puddler: A peddler containing of a wooden plank provided with number of small spikes on one edge and hang at the center.

Plantation spraying: The spraying performed vertically to cover the height of the plant or trees.

Planter: A machine used for precision drilling, hill dropping or check row planting.

Plaster of paris: Calcined gypsum, marketed in the form of a white powder. When mixed with water, it sets quickly, and is useful in making casts and models.

Plastering trowel: A thin rectangular piece of steel 10-12 cm. wide and 25-30 cm. long, with handle attached, offset but parallel to the blade.

Plastic: A substance capable of being molded or modeled. In the broad sense there are many plastic substances, from pitch to Portland cement, and on, through the synthetics. All flow at some stage of use. There are many combinations of plastic source materials; therefore, there is a large and growing number of members in the plastics family.

Plastic wood: A wood compound, which quickly hardens on exposure to air, for filling in cracks and defects. It can be painted over almost immediately after applying.

Plate feed mechanism: A feed mechanism in which cells are provided on the periphery of a horizontal or slightly inclined plate.

Plate and flicker mechanism: A combination of horizontal, rotating plate feed mechanism and a rotating flicker distributing mechanism.

Plate and flicker type distributor: gravity - feed fertilizer distributor in which fertilizer is flicked from plate mounted horizontally below the bottom of the hopper by means of a rotating shaft fitted with fingers.

Plate clutch: A clutch, which transmits power through two or more plates, which are held in contact by the pressure of springs. *Refer* Fig. 15C.

Plate current: The pulsating direct current which flows in the plate circuit of any stage.

Plate glass: A high-grade glass cast in the form of a plate or sheet and subsequently polished; usually thicker than window glass, of smoother surface and better quality.

Platform: A horizontal structure usually covered with wood or metal and set on uprights to form an elevated flooring or stand.

Play (mechanism)**:** The motion between poorly fitted or worn parts.

Plenum: An air chamber maintained under pressure (positive or negative) usually connected to one or more disturbing ducts in a drying or aeration system.

Pliers: The lowest square-shaped part of the base of a column or pedestal.

Plough: An implement used to cut the furrow slice with partial or complete soil inversion and in some cases to break it. *Refer* Fig. 5R.

Plough bottom (body)**:** An assembly comprising the frog, standard (leg), mouldboard, share and landside.

Ploughing: A primary tillage operation, which is performed to cut, breaks, and invert the soil partially or completely.

Plug (electric)**:** A device with conducting projections, which fit into slots, making electrical contact between an appliance and a source of supply.

Plug valve: A valve fitted with a plug that has a hole through which fluid flows and that is rotatable through 90^0 for operation in the open or closed position. Also known as plug cock.

Plumb: To test or true up vertically, as a wall by means of a plumb line.

Plumbing: Installation and repair of water pipes, tanks, bathroom fixtures, sewage lines, etc.

Plumb and level: A piece of well-finished hardwood or metal with bubble set crosswise for testing vertical accuracy.

Plumb bob: The weight used at the end of a plumb line.

Plunger pump: A reciprocating pump where the packing is on the stationary casing instead of the moving piston.

Pneumatic agitation: Agitation of the spray mixture, dust or granules inside the tank or hopper using airflow.

Pneumatic agitator: A device used for pneumatic agitation (of seed, fertilizer etc in the hopper).

Pneumatic brakes: Brakes operated either by air pressure or vacuum.

Pneumatic distributor: A fertilizer distributor in which the fertilizer is transported either to a distributor mechanism or directly to the soil surface by pneumatic means.

Pneumatic drill: Compressed-air drill worked by reciprocating piston, hammer action, or turbo drive.

Pneumatic drilling: Drilling a hole when using air or gas in lieu of conventional drilling fluid as the circulating medium; an adaptation of rotary drilling.

Pneumatic duster: An appliance used for pneumatic dusting.

Pneumatic dusting: Distribution of dust by means of flow of gas, usually air.

Pneumatic governor: It is used on both carburetor and diesel engine. In this governor, a pressure depression created at the venturi of intake manifold moves the diaphragm, which in turn operates a lever connected to the fuel injection pump in diesel engine or to the connecting lever of the throttle valve in carburetor engine.

Pneumatic riveter: A riveting machine having a rapidly reciprocating piston driven by compressed air.

Pneumatic seed drill: A seed drill in which the metering of seeds is done by air stream at set spacing in continuous parallel rows.

Pneumatic-spacing drill with blower: A spacing drill in which distribution is effected by means of an air jet directed at the feed mechanism, which leaves only single seeds in the cells of the distributor mechanism. A seed spacing is usually controlled either by changing the speed of rotation or the spacing of the perforations of alternative seeding discs.

Pneumatic-spacing drill with perforated discs: A spacing drill in which the distribution is affected by providing a negative pressure to the metering mechanism which consists of a perforated disc. A seed spacing is usually controlled either by changing the speed of rotation of the disc, or by selecting discs having different whole spacing. Perforated drum or rim may be used instead of perforated disc.

Pneumatic sprayer: A sprayer used for pneumatic spraying.

Pneumatic spraying: Spraying performed by the action of flow of gas.

Pneumatic tyre: A tyre usually consisting of an outer casing or shoe and an inner tube which, when the tire to retain its shape under load.

Pneumatic tools: Tools operated by air pressure.

Pneumatic tyred wheel: A wheel fitted with a pneumatic tyre.

Pneumatic wheel vehicle: Animal drawn vehicle having pneumatic wheels.

Pod: Unbroken shell with kernel inside.

Polarity: The condition in an electric component or circuit that indicates the direction of current flow. The identification of one point as positive (+) and another point as negative (-) for a voltage.

Polarization: In a primary cell, the collection of hydrogen bubbles on the positive plate, which increases the internal resistance, and diminishes the current strength.

Poles: The positive and negative terminals of an electric circuit.

Pole shoes: The end part of a generator field pole, ordinarily curved approximately the cylindrical shaped of the armature and designed to hold the field coils in position.

Polishing (rice)**:** Removal of bran layer from the brown rice by mechanical operations. It is also known as whitening or pearling.

Pollutant: Any substance that adds to the contamination or degrading of the environment. In a vehicle, any substance in the exhaust gas from the engine or evaporating from the fuel system.

Pollution: Any gas or substance that makes the environment less fit. Types of pollution include: air, ground water, ocean, noise, etc.

Polygon: A plane figure of many sides and angle, especially more than four.

Polygon of forces: An expansion of the triangle of forces. If any numbers of forces are represented in magnitude and direction by the sides of a polygon taken in order, the resultant is the closing side of polygon.

Polytropic process: A thermodynamic process, in which pressure and volume are often related by $PV^n = C$, where n and C are constants, during expansion and compression processes of real gases.

Poppet: 1. The headstock of a lathe. 2. A lathe center.

Poppet valve: A cam operated or spring-loaded reciprocating engine mushroom type valve used for control of admission and exhaust of working fluid; the direction of movement is at right angle to the plane of its seat. *See* Fig.4P.

Porcelain: The common main of the spark plug insulator.

Pore hal: A narrow wooden plough with sowing attachment.

Port: An opening through which fuel may be admitted to the combustion chamber of the engine or one from which exhaust gases are released.

Port (fuel) delivery systems: This system forms the fuel-air mixture during the intake stroke. It is injected at the inlet port.

Positive chamber: The condition existing when the two tyres are closer together at the bottom than they are at the top.

Positive displacement pump: A pump in which a measured quantity of liquid is entrapped in a space, its pressure is raised, and then it is delivered; for example, a reciprocating piston-cylinder or rotary-vane, gear, or lobe mechanism.

Positive feed: Direct feed motion by means of gears, without friction clutches or belts.

Positive plate: Any one of the plates used to form the positive group for the storage battery element; presumed to be the group of plates from which the electrical current flows.

Potato digger: A machine to bring the potatoes to the surface of the soil for collection.

Potato digger shaker: A potato digger to separate potatoes from soil by means of reciprocating grids and to deliver the potatoes in a narrow band on the ground.

Potato elevator digger: A potato digger to separate potatoes from soil by means of a open web elevator and to deliver the potatoes in a narrow band on the ground.

Potato grader: A machine used to grade potatoes according to size.

Potato harvester: A potato digger to separate potatoes from soil, trash, etc, and to deliver them into a container.

Potato plough: A plough used for harvesting potatoes and also bringing the potatoes to the surface of the soil.

Potato spinner: A potato digger to deliver potatoes to one side by means of fingers carried on a rotating hub.

Potential: 1.An electrical state or the electrical state of a point. It is measured by the work done in bringing a unit charge from an infinite distance up to the point. It is measured in practical units called volts. 2. A condition at a point in space, due to local attraction or repulsion, such that a mass, electric charge, etc., at such point becomes capable of doing work.

Potential difference: The difference in the electrical state of two points, measured in volts.

Potential energy: The energy that a system possesses as a result of its elevation in a gravitational field with reference to an arbitrary chosen datum and is expressed as PE = mgz, m is mass, g is gravitational acceleration and z elevation.

Pound-force (lbf): In the English system, is the force unit defined as the force required to accelerate a mass of 32.174 lbm (1 slug) at a rate of 1 ft/s^2.

Power: The rate at which work is performed or the amount of work done in a unit of time. In mechanics power is measured: force x distance ÷ time. It is expressed in kilocalorie per minute or second. The unit of electric power is the watt, in which is measured the rate of work done by electricity in motion.

Power brakes: Brakes operated hydraulically through a booster system.

Power clutch: A type of electromagnetic disk clutch in which the space between the clutch members is filled with dry, finely divided magnetic particles; application of a magnetic field coalesces the particles, creating friction forces between clutch members.

Power factor: The ratio of the true power (watts as read by a wattmeter) to the apparent power (volts times amperes as read by a voltmeter and ammeter respectively)

Power feed: The automatic feed of a lathe, planer, screw cutting, or other machine.

Power input connection shield: A rigid guard fitted on a machine, shielding the power input connection.

Power lawn mower: A grass-cutting machine, which is powered by mechanical means or electric motor.

Power lift: A mechanism driven by a tractor's power unit to raise or lower mounted or semi-mounted equipment.

Power operated mower: A mower operated by tractor or power tillers.

Power operated: Operated by a prime mover, such as engine, tractor, power tiller and electric motor.

Power operated winnowing fan: A winnowing fan operated by prime mover, such as engine, electrical motor, tractor, power tiller.

Power outlet: Any outlet, which transmit the engine power to the tractor in order to make it functional, such as PTO, belt pulley and drawbar.

Power output: For an engine power output is based on the speed the engine is turning, multiplied by the torque produced.

Power steering: Hydraulically operated through a booster built into the steering unit attached to the steering linkage. Permits free steering when the engine is running.

Power stroke: 1. The arc of crankshaft travel while the pressure of ignited gases is exerted on the piston. 2. The piston stroke on which the engine is delivering power. *Refer* Fig. 9F.

Power take off (PTO): A shaft, usually externally splined, to transmit torsional power to another machine. It is abbreviated as PTO.

Power take off cover: A protective cover enclosing a power takes off.

Power take off drive shaft: The assembly from power take off yoke boss to the power input connection yoke boss.

Power take off power: Power obtained at the main power take off with the governor control in position recommended by the tractor manufacturer for PTO work, the tractor being stationary.

Power take off shield: A rigid guard fitted on a tractor, covering a tractor power takes off as a safety device.

Power thresher: A machine used for threshing, operated by a prime mover, such as electric motor, tractor engine, power tiller etc.

Power tiller: A prime mover, operated by a person walking behind it, performs field operation or used for transporting purpose. It is also known as hand or walking type tractor. Power tiller may be provided with attachment of riding.

Power transformer: A device for connecting power from a high voltage and low current to a low voltage and high current, or vice versa, the amount of power so changed remaining theoretically the same. The frequency in which these are used is usually 60 cycles.

Power unit: The watt is the unit of power in electrical circuits. The product of voltage times the amperage gives the wattage.

Ppb: Parts per billion

Pre cleaner: A separator to remove extraneous matters from grain before drying by means of sieves and air stream.

Precision drilling: Uniform placing of seeds in rows at a pre-determined depth and seed rate.

Precision grinding: Machine grinding in which the tolerances are exceedingly close.

Precision lathe: A small bench lathe used for very accurate work.

Precooler: A device for reducing the temperature of a working fluid before a machine uses it.

Pre-delivery set-up publication: Documents, which outline, in detail, the procedures for properly preparing equipment for delivery to the customer.

Preheater: A device for preliminary heating of a material, substance, or fluid that will undergo further use or treatment by heating.

Preignition (premature ignition): Ignition occurring earlier than intended. For example, the explosive mixture being fired in a cylinder as by a flake of an incandescent carbon before the electric spark occurs. This results in an inefficient, rough running engine.

Press fit: A fitting together of parts by pressure; slightly tighter than a sliding fit. Also known as the Forced fit or Drive fit. This term is used when the shaft is slightly larger than the hole must be forced into place.

Press wheel: A wheel, which compacts soil and covers seeds in the furrow. *See* Fig.6P.

Pressing worm: A steel worm with a helical thread on its length equal to one revolution and mounted on the main shaft in the chamber. It pushes the feed ahead towards the cone head. The setting of the worm assembly differs with the expeller and the type of seed to be crushed.

Pressure: 1. The exertion of continuous force on or against a body by another in contact with it, expressed as force per unit area. 2. Electromotive force (e.m.f.) commonly spoken of as voltage

Pressure angle: The angle that the line of force makes with a line at right angles to the centreline of two gears at the pitch points.

Pressure bar: 1. A spring or weight load bar, which assists in penetration of the furrow opener. 2. A bar that holds the edge of a metal sheet during press operations, such as punching, stamping, or forming, and prevents the sheet from buckling or becoming crimped.

Pressure chamber: A chamber with or without air pressurization to even out the fluctuations of the liquid pressure and induce uniform flow of liquid. It is also known as air vessel, air bottle or air chamber.

Pressure differential : The difference between the pressure of the air/fuel mixture in the intake manifold and atmospheric pressure.

Pressure fraction: The ratio of the component pressure to the mixture pressure. Note that for an ideal-gas mixture, the mole fraction, the pressure fraction, and the volume fraction of a component are identical.

Pressure plate: That part of the friction clutch, which is thrust against the driven plate by means of spring pressure. It is mounted on and rotates with the flywheel. Refer Fig. 15C.

Pressure pulsation damper: A device for reducing pressure pulsations.

Pressure ratio: The ratio of final to initial pressures during a compression process.

Pressure recovery factor: A measure of a diffuser's ability to increase the pressure of the fluid, is expressed in terms of the ratio of the actual stagnation pressure of a fluid at the diffuser exit relative to the maximum possible stagnation pressure.

Pressure regulator: An automatic device to control the pressure of a liquid or gas within a range of settings. The volume of gas may vary.

Pressure relief device (PRD): A safety device on a pressure vessel, actuated by either overpressure or over temperature that relieves the pressure by opening to vent some or all of the contents of the vessel.

Pressure relief valve: A valve designated as a safety device to open and remain open, to discharge a fluid whenever the fluid pressure reaches the start to discharge setting of the valve. When the fluid pressure drops somewhat below this setting, the relief valve automatically closes.

Pressure system for fuel supply: The system used before the invention of the vacuum tank, for supplying gas when the tank was lower than the carburetor. A small hand pump is used to provide initial pressure. The engine, when running, operates a small pump thus maintaining pressure. In this system the gas tank must not be vented.

Pressure transducer: A sensor that converts pressure readings to electrical signals. They are made of semiconductor materials such as silicon and convert the pressure effect to an electrical effect such as a change in voltage, resistance, or capacitance. Pressure transducers are smaller and faster, and they are more sensitive, reliable, and precise than their mechanical counterparts.

Pressurize: To apply more than atmospheric pressure to gas or liquid.

Preventive maintenance: Systematic series of inspections and operations performed periodically to maintain or improve the efficiency and performance of the machine.

Primary air: That portion of the combustion air introduces with the fuel in a burner.

Primary battery of cell: The ordinary dry cell or a battery of dry cells; a primary cell is used up in producing electric current and cannot be recharged.

Primary cell: A device for transforming chemical energy to electrical energy. It consists essentially of jar containing the solution, or electrolyte, and two plates of electrodes.

Primary coil: The coil into which the original energy is introduced and which sets up magnetic lines of force to link with another coil in which energy is induced. Refer Fig. 6B.

Primary or fundamental dimensions: Dimensions such as mass m, length l, time t, and temperature *T*, are the basis for the derivation of secondary dimensions.

Primary shear surfaces: Initial and distinct shear surfaces, which appear during failure and are caused mainly by soil movement resulting from the advance of a tool.

Primary shoe: The shoe, which first receives the breaking action from the brake drum and carries it forward against the secondary shoe.

Primary terminal: Any terminal of a six-volt or low-tension circuit.

Primary tillage: Tillage operations, which constitute the initial major soil working operations. It is normally designed to reduce soil strength, cover plant materials and re-arrange aggregates.

Primer: Device for supplying extra fuel to aid starting.

Principle of moments: The algebraic sum of any number of forces with respect to a point equals the moment of their resultant about that point.

Process: Any change that a system undergoes from one equilibrium state to another. To describe a process completely, one should specify the initial and final states of the process, as well as the path it follows, and the interactions with the surroundings.

Process chart: A graphical presentation of events occurring during a series of actions or operations and of information pertaining to their events.

Production engineer: One who is responsible for the maintenance of production. He also directs tooling operation and the design of fixtures and appliances to secure the most efficient manufacturing methods.

Products (combustion)**:** The components that exit after the reaction in a combustion process.

Projection: 1. A jutting out, a prominence. 2. In drawing, the method by which one or more views of an object are used as an aid in securing additional views.

Projection welding: A resistance welding process wherein localization of heat between the end of one member and the surface of another is effected by projections.

Projector: A device for projecting a beam of light, as a searchlight projector.

Propane (C_3H_8): A type of liquid petroleum gas (LPG) that is liquid below -42°C at atmospheric pressure. Propane gas is heavier than air.

Propeller: The drive shaft turned by the universal joint at the transmission end and delivering power to other toed implements.

Propeller efficiency: The ratio of thrust power to power input of a propeller.

Propeller shaft: Also called drive shaft. It is the shaft that delivers the power from the transmission to the rear axle.

Propeller thrust: The component parallel to the propeller axis of the total air force on the propeller.

Property: Any characteristic of a system. Some familiar properties are pressure P, temperature T, volume V, and mass m. The list can be extended to include less familiar ones such as viscosity, thermal conductivity, modulus of elasticity, thermal expansion coefficient, electric resistivity, and even velocity and elevation.

Proton: A subatomic particle in the nucleus of an atom that carries a positive electric charge, and is not moveable by electrical means.

Protractor: An instrument for measuring and lying off angles on paper, used in drawing and plotting. *See* Fig.5P.

Pruning: Removing live or dead braches as or multiple leaders from standing tree for the improvement of the tree or its timber.

Pruning knife: A folding knife with handle used for pruning of small plants.

Pruning saw: A metal strip with serrated teeth and handle for cutting green stock.

Psi: Pounds per square inch; a unit of pressure. Psi is more correctly indicated as psia-pounds per square inch absolute, or the absolute pressure measured relative to a perfect vacuum. Related to this is psig -pounds per square inch gauge as measured relative to atmospheric pressure.

Psychometric chart: Presents the properties of atmospheric air at a specified pressure and two independent intensive properties. The psychometric chart is a plot of absolute humidity versus dry-bulb temperature and shows lines of constant relative humidity, wet-bulb temperature, specific volume, and enthalpy for the atmospheric air.

Puddle: To settle loose dirt by application of water. The batch of molten iron in the puddling furnace.

Puddler: The implement used to churning the soil in standing water.

Puddling: The mechanical manipulation of soil in presence of standing water in the field to create and impervious hard pan below the puddle zone so as to prevent loss of water through leaching and facilitate transplanting of paddy seedling by making the soil softer

Pull: The total force required to pull an implement.

Puller: Any mechanical or hydraulic device for removing, by pulling action, parts which are tightly fitted; e.g., wheel puller or gear puller.

Pulley: 1. A metal wheel with a V-shaped groove around the rim that drives, or is driven by, a belt or rope. 2. A wheel used to transmit or receive power through a belt, which travels over its face.

Pulley block: A sheave pulley or series of such pulleys, enclosed between metal or wooden side plates which carry the shaft or pin on which the pulleys revolve.

Pulley lathe: A lathe used for turning either a straight or crowned face on pulleys.

Pulley top: A top with a very long shank, used for tapping setscrew holes in the hubs of pulleys.

Pull-in torque: The largest steady torque with which a motor will attain normal speed after accelerating from a standstill.

Pulverization: The general fragmentation of a soil mass resulting from the action of the tillage forces.

Pulverization roller: A helically formed roller attachment behind a cultivator for breaking the clods.

Pump: A device used for lifting or forcing liquids, either by means of a bucket or of a piston working in a closed cylinder

Pump output: Volume of liquid discharged by a pump per unit time.

Punch: A shearing tool made of steel, used to remove material whose shape is the same as that of the punch. The perforate or cut with a punch, as opposed to drilling or boring. *Refer* Fig.9C.

Punching: The making of holes through plates by means of a punching machine, as on boilerplate.

Pure substance: A substance that has a fixed chemical composition throughout.

Purge: The use of a gas to flush residual gases and/or liquids from a container.

Purlins: Timbers spanning from truss to truss, and supporting the rafters of a roof.

Push button starter: The push button on the instrument panel that operates to close an electrical circuit which, in turn, causes solenoid switch to close the starting motor circuit.

Push rod: A connecting link in an operating mechanism, such as the rod interposes between the valve lifter and rocker arm on an overhead valve engine.

Putty: A composition of whiting and linseed oil, used for filling small holes in woodwork, and securing panes of glass in sash.

P-v-T surface: A three-dimensional surface in space, which represents the P-v-T behavior of a substance. All states along the path of a quasi-equilibrium process lie on the P-v-T surface since such a process must pass through equilibrium states. The single-phase regions appear as curved surfaces on the P-v-T surface, and the two-phase regions as surfaces perpendicular to the P-T plane.

Pyrolysis: The chemical decomposition brought about by heat.

Pyrometer: An instrument for measuring very high degrees of heat, as in furnace.

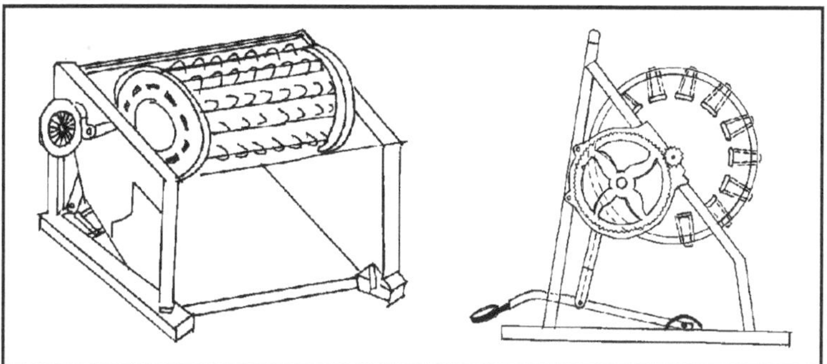

Fig. 1P Paddy thresher (Pedal operated)

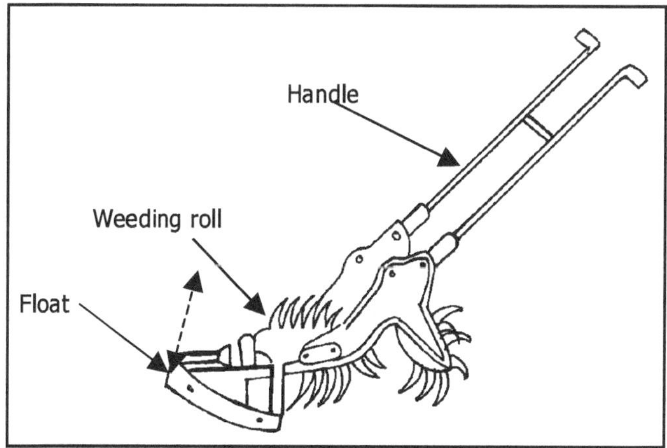

Fig. 2P Paddy weeder

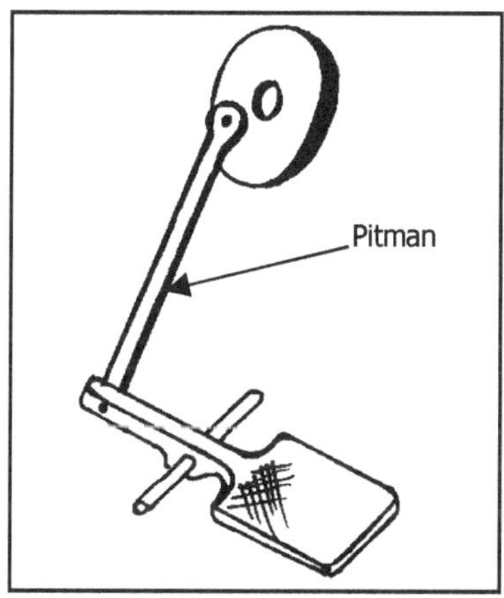

Fig. 3P Pitman

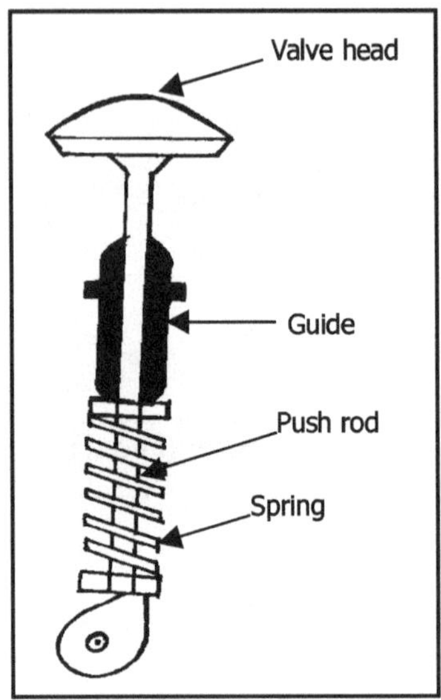

Fig. 4P Poppet valve

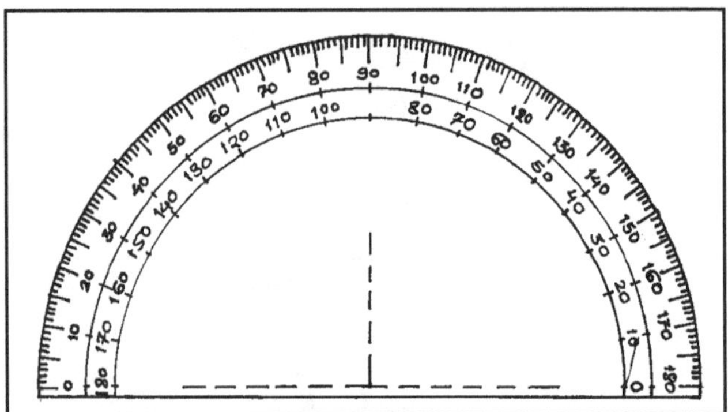

Fig. 5P Protractor

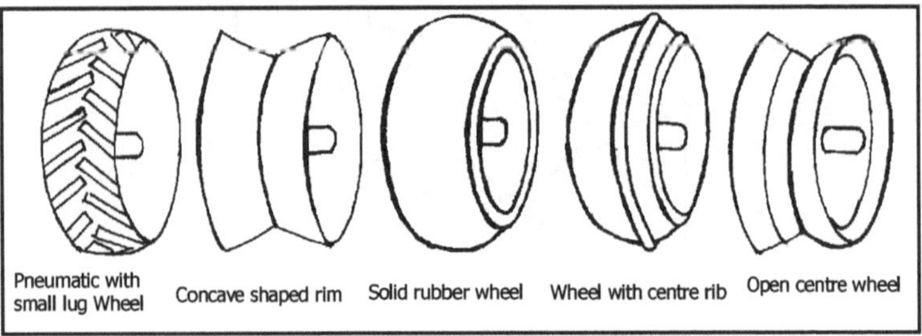

Fig. 6P Types of Press wheel

Quadrangle: A square or quadrangular space surrounded by buildings, as on college grounds.

Quadrant: 1. The quarter of a circle, an arc of 90^0. 2. An instrument for measuring altitudes.

Quadrilateral: A plane figure with four sides and four angles.

Quarter turn belt: The arrangement of a belt to drive two shafts, which are at right angle.

Quarter turn drive: A belt drive connecting pulleys whose axes are at right angles.

Quartering machine: A machine that bores parallel holes simultaneously in such a way that the centre lines of adjacent holes are at 90 deg. apart (axes of the holes are perpendicular to the plane on which an angular measurement are taken.)

Quartz: A hard, crystalline mineral occurring as a rock (SiO_2). It is usually colorless, but is often colored by impurities.

Quasi-static, or quasi-equilibrium, process: A process which proceeds in such a manner that the system remains infinitesimally close to an equilibrium state at all times. A quasi-equilibrium process can be viewed as a sufficiently slow process that allows the system to adjust itself internally so that properties in one part of the system do not change faster than those at other parts.

Quench: The removal of heat from the end gas or outside layers of air-fuel mixture during combustion, by the cooler metallic surfaces of the combustion chamber, thus reducing the tendency for combustion to occur.

Quenching: The process of dipping of heated steel into water, oil, or other bath, to impart necessary hardness.

Quench area: The area of the combustion chamber near the cylinder walls that tends to cool (quench) combustion because of the nearby cool water jackets.

Quenching oils: Oils used in heat-treating. Fish oils are much used but have offensive odors. Mineral, fish, vegetable, and animal oils are often compounded and sold under trade names.

Quick acting shut-off valve: A device enabling a spray system to be opened or closed instantly.

Quick change: The arrangement of gears on a lathe in such a manner as to permit change of feed by shifting levers instead of removing and replacing gears.

Quick release hydraulic coupling: A self-sealing coupling providing quick connection and disconnection of a tractor to implement hydraulic pipeline.

Quick release mechanism: Device enabling the operator to free himself quickly and avoid an accident in case of emergency.

Quick return: A term applied to shapers, planers, and other metalworking machines, which are so constructed that the return stroke is such more rapid than the forward or cutting stroke.

Quick-change gearbox: A cluster of gears on a machine tool, the arrangement of which allows for the rapid change of gear ratios.

Quoins: Large squared stones set at the angles of buildings, buttresses, etc. A sedge or pair of wedges used in locking up type in a chase or gallery.

—□—□—

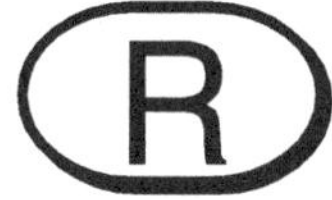

Rabbet: A rectangular groove or cut made in the edge of a board, so that another similar cut may fit into it to form a rabbet joint.

Rack and pinion: A gear arrangement consisting of a toothed bar that meshes with a pinion. These are used to transform circular motion in to rectilinear motion and vice versa.

Rack and pinion steering: A steering system in which the rotation of pinion gear at the end of the steering column moves a toothed bar (the rack) left or right to transmit steering movements.

Radial arm: The movable cantilever, which supports the drilling saddle in a radial drilling machine.

Radial bearing: Bearing made with cylindrical or with barrel shaped rollers. The direction of action of the load transmitted to the axis of shaft.

Radial drill: 1.A drilling machine in which the drill spindle can be moved along a horizontal arm, which itself can be rotated about a vertical pillar. 2. A heavy drilling machine, so constructed that the position of the drill can be adjusted to the work without moving the latter.

Radial engine: An engine, which has stationary cylinders, arranged radially around a common crankshaft.

Radial force: In machining, the force acting on the cutting tool in a direction opposite to depth of cut.

Radial motion: Motion in which a body moves along a line connecting it with an observer or reference point. For example, the motion of stars; they move towards or away from the earth without change in apparent position.

Radial saw: A power saw that has a circular blade suspended from a transverse head mounted on rotable over arm.

Radiation: 1. The transfer of energy due to the emission of electromagnetic waves (or photons). 2. The emission of very fast atomic particles or rays by nuclei.

Radiator: Device employed to dissipate excess heat from an internal comb-engine, consists of large number of this walled tubes through which cooling water, passes around the cylinder and combustion spaces, is circulated by thermosiphon action or by a pump. *See* Fig.1R.

Radiator hose: The hose, which connects radiator and engine and conveys the cold water from radiator to engine surrounding.

Radiator pressure cap: A cap placed on the radiator filler tube that allows for the pressurization of the cooling system for more efficient operation.

Radiator temperature drop: In internal combustion engines, the difference in temperature of the coolant liquid entering and leaving the radiator.

Radio: A preferred name for wireless. Often incorrectly used where a compound word would be more proper, as radiotelephone, radiotelegraph etc.

Radius cutter: A formed milling cutter with teeth ground to produce a radius on the work piece.

Radius gauge: An instrument for measuring the radii of fillets and rounded corners.

Radius of gyration: Equals the square root of the quotient of the moment of inertia divided by the area of the section. $R = v (I / A)$. Where, R = radius of gyration, I = Moment of inertia, and A = area.

Radius rod: A brace member run in between the axle housing and the front end of the transmission box.

Rake: A term usually applied to signify the angles of metal turning tools, as side rake, front rake, etc.

Ramjet engine: A properly shaped duct with no compressor or turbine, and is sometimes used for high-speed propulsion of missiles and aircraft. The pressure rise in the engine is provided by the ram effect of the incoming high-speed air being rammed against a barrier. Therefore, a ramjet engine needs to be brought to a sufficiently high speed by an external source before it can be fired.

Ramming: The effect obtained when the air intake to the engine is placed in the slip steam in such a manner as to take advantage of the difference in velocity of this air intake and the slipstream, in order to increase the pressure in the induction system.

Rankine cycle: An ideal thermodynamic cycle consisting of heat addition at constant pressure, isentropic expansion, heat rejection at constant pressure, and isentropic compression.

Rankine cycle with reheat: It is a modification of the Rankine cycle in which the steam is expanded in the turbine in two stages and reheated in between. Reheating is a practical solution to the excessive moisture problem in the lower-pressure stages of turbines, and it is used frequently in modern steam power plants.

Rankine efficiency: The efficiency of an ideal engine operating on the Rankine cycle under specified conditions of steam temperature and pressure.

Rankine scale: Named after William Rankine (1820-1872) is the thermodynamic temperature scale in the English system. The temperature unit on this scale is the Rankine, which is designated by R.

Rapeseed oil: A heavy, brown oil obtained from rape seed. Used as a lubricant, and in the heat treatment of steel.

Rasp bar cylinder type thresher: A thresher, the threshing unit of which consists of bars with serrations having an open concave.

Rasp drum: A cylinder having bars with serrations.

Rasp: A file like tool having coarse projections for abrasion.

Ratchet: A gear with triangular shaped teeth, adapted to be engaged a pawl, which either imparts intermittent motion to the ratchet, or locks it against backward movement when operated otherwise. *See* Fig.2R.

Rated capacity: The maximum capacity for which a boiler is designed, measured in pounds of steam per hour delivered at specified conditions of pressure and temperature.

Rated engine speed: The rotational speed of an engine specified as the allowable maximum for continuous reliable performance.

Rated horsepower: The maximum, allowable, continuous power output of an engine, turbine motor, or other prime mover. Valve used by the engine manufacturer to rate the power of the engine for safe loads.

Ratio: The relation or proportion of one number or quantity to another. Usually expressed as a numerical relationship, as in 2:1.

Reach: The horizontal distance measured perpendicular to the longitudinal centre line of the combine from the innermost point of the unloader discharge opening to the outermost point of the header on the unloader side expressed in millimeters.

Reactant: 1. A substance that enters into chemical combination with another substance. 2. A component that exist before the reaction in a combustion process.

Ream: To finish a hole accurately with a rotating fluted tool.

Reaming: The process of smoothing the surface of holes with a reamer.

Reaper: A machine to cut grain crops.

Reaper binder: A reaper, which after cutting the crops ties them into neat and uniform sheaves.

Reaping attachment: A set of attachments to adopt a mower to cut grain crops.

Rear axle: More properly the rear system; consists of axle housing, the gears, the axle shaft drive, and other essential parts (used in vehicles).

Rear beater: An element placed on the rear side of the cylinder and above to reward of concave or concave grate extension to transition grates to assist the deflection of straw on straw walker. It also assists in stripping the straw from the cylinder and prevents it from wrapping around.

Rear furrow wheel: The rear wheel of a plough, which runs in the furrow.

Rear handle: Support handle located or towards the rear of engine housing.

Rebound stroke: The motion of the shock absorber as the springs of the car tend to return to normal position after encountering a depression or obstruction on the highway.

Receiver: The instrument, which is held to the ear in receiving a telephone message. It consists of a hard-rubber container or shell, which holds a permanent horseshoe or U magnet with a thin iron diaphragm close to its poles.

Reciprocating motion: A back and forth movement, such as the action of a piston in a cylinder.

Reciprocating mower: A mower having knife section that reciprocates against the stationery fingers (guards).

Reciprocating plate distributor: A gravity-feed fertilizer distributor whose discharge mechanism consists of reciprocating plates in the base of the hopper.

Reciprocating plate mechanism: A combined feed and distributing mechanism at the base of a hopper consisting of a series of reciprocating superimposed slotted or perforated plates, each pair being separated by a fixed plate or other components. *See* Fig.3R.

Reciprocating power harrow: Harrow fitted with rigid tynes driven by power take-off in a reciprocating transverse or rotary motion as the machine travels forward.

Recirculating balls (steering gears) : A train of balls in a steering gear so arranged as to travel in the sliding nut over the worm shaft, then outlet through a tubular conductor, and back into the nut. Used to promote ease of steering.

Recording thermometer: A thermometer, which makes a permanent record of variations in temperature usually by an inked line traced automatically on a paper disk or roll.

Rectifier: Any device used to change alternating current into direct current. A rectifier causes succeeding waves of the current to flow in the same direction through the electric circuit.

Reduced pressure (PR) : The ratio of the pressure to the critical pressure.

Reduced temperature (TR) : The ratio of the temperature to the critical temperature.

Reduced tillage: A tillage system in which the primary tillage operation is performed in conjunctions with special planting procedure in order to reduce or eliminate secondary tillage operation.

Reducer: Any one of the various pipe connections so constructed as to permit the joining of pipes of different sizes, such as reducing sleeve, reducing elbow reducing tee, etc.

Reduction gears: Gears used on a shaft to reduce the speed.

Redwood: A giant tree of California. The wood is soft, reddish in color, and light in weight.

Reel: Revolving slats or arms with battens arranged parallel to the cutter bar to hold the crop being cut by the knife and to push and guide it to a conveyor platform or feeder conveyor auger. The reels may be of spring tine type or slat type.

Reference state (thermodynamics) : The state chosen to assign a value of zero for a convenient property or properties at that state.

Reformat: The hydrogen rich gas mixture that results from processing hydrogen-containing fuels in a reformer.

Reforming: A chemical process that reacts hydrogen-containing fuels in the presence of steam, oxygen, or both into a hydrogen rich gas stream.

Refrigerant: The working fluid used in the refrigeration cycle.

Refrigerator: 1. An insulated, cooled compartment. 2. A cyclic device, which causes the transfer of heat from a low-temperature region to a high-temperature region. The objective of a refrigerator is to maintain the refrigerated space at a low temperature by removing heat from it.

Refrigerator coefficient of performance: Efficiency of a refrigerator, expressed as desired output divided by required input.

Regeneration: A process during which heat is transferred to a thermal energy storage device (called a regenerator) during one part of the cycle and is transferred back to the working fluid during another part of the cycle.

Regenerative (or dynamic) **braking:** An energy recovery system in which energy generated by vehicle braking is converting into electricity and stored.

Registration of mower: A mower is said to be in proper registration, when the knife section stops in the centre of its guard on every stroke. *See* Fig. 4R.

Regulator: A device that controls generator output to prevent excessive voltage, excessive current output, or excessive air/gas pressure.

Reinforcement: Embedment within a plastic sheet or form which give the product strength, toughness, and rigidity, Reinforcement refers principally to laminated plastic, where the chief ingredient is glass fiber, either in cloth or mat form, and in certain applications, cut-glass-fiber is the added material. Glass fiber lends great strength to the product, but other material such as resin fiber, sisal, asbestos, and others, are also employed.

Relative humidity: A measure of the amount of moisture the air holds relative to the maximum amount the air can hold at the same temperature. The relative humidity can be expressed as the ratio of the vapour pressure to the saturation pressure of water at that temperature.

Relay: An electric switch operated magnetically, which open or close a local circuit under given conditions in the main circuit.

Reliability : It is the probability of an item or component part that it will perform a required function under specified condition for a state period of time.

Relief valve: A valve that opens when a preset pressure is reached. This relieves or prevents excessive pressures.

Relining brakes: Applying new brake lining material to the brake shoes or brake bands.

Reluctance: The inability of a metal to magnetic lines of force.

Remote control: Control from a distance, especially b means of electricity or electronics; a controlling switch, lever, or other device used in this kind of control. The control of electrical apparatus or machinery from a distance by the use of a relay or other electromagnetic device.

Remove and replace (R and R): To perform a series of servicing procedures on an original part or assembly; includes removal, inspection, lubrication, all necessary adjustments, and reinstallation.

Replace: To remove a used part or assembly and install a new or rebuilt part or assembly in its place; includes cleaning, lubricating, and adjusting as required.

Replaceable heel: A heel of the landside, which is replaceable.

Replica: An exact copy or production.

Repulsion: The action of a force by which two similarly charged bodies tend to repel each other.

Residual magnetism: Magnetism remaining in an electromagnet core after the producing current has been turned off.

Residue: Foreign material including roots, remaining in or on the soil in significant amounts such that they can be detected, or they can exert a separate and distinct influence on soil properties or machinery operations.

Residue processing: Operation that crush, cut, anchor or otherwise handle residue in conjunction with soil manipulation.

Resilience: The act or power of springing back; capability of a strained body to recover its size and shape after deformation, or the flexing (bending) action of the springs. The more resilient spring gives a softer ride; a less resilient spring gives a harder ride but is capable of carrying a heavier load.

Resistance (R) : The characteristic of a device to oppose the passage of electrical current. The opposition to a flow of current through an electric circuit or device; measured in ohms. A voltage of 1 volt will cause 1 ampere to flow through a resistance of 1 ohm.

Resistance box: A box containing known resistances, the amount of which can be varied by means of plugs or dials.

Resistance butt-welding: A group of resistance-welding processes wherein the fusion occurs over the entire cross-sectional area.

Resistance coil: A coil of wire of high specific resistance, such as Nichrome or iron, inserted in a circuit to decrease the current flow.

Resistance unit: Usually a small coil of electrical resistance wire wound to a certain capacity and used to control the flow of electrical current.

Resistance welding: A pressure-welding process wherein the heat is obtained from the resistance to the flow of an electric current.

Resistive temperature device (RTD): A device whose electrical resistance changes in proportion to its temperature.

Resonance: A condition reached in an electrical circuit when the inductive reactance just neutralizes the capacitance reactance leaving Ohmic resistance as the only opposition to the flow of current.

Retard: The act of setting the time of the spark occurrence so as to have it comes at a later point with reference to the travel of the piston in the cylinder.

Retracting spring: Springs which pull the brake shoes away from the brake drum in hydraulic brakes.

Reversed Carnot cycle: A reversible cycle in which all four processes that comprise the Carnot cycle are reversed during operation. Reversing the cycle will also reverse the directions of any heat and work interactions. The result is a cycle that operates in the counterclockwise direction and becomes Carnot refrigeration cycle.

Reversible plough: A plough, which turns the furrow slice to the right and to the left alternatively with respect to the forward travel. *See* Fig.5R.

Reversible plough (half turn type)**:** A plough with bottoms (bodies) mounted on opposite sides of a beam or frame, which can be rotated approximately 180^0 about a longitudinal axis.

Reversible plough (quarter-turn type)**:** A plough of which the bottoms (bodies) mounted on the same beam are fixed at an angle of 90^0 with each other and able to pivot through a quarter circle about the longitudinal axis.

Reversible plough (roll-over type)**:** A plough of which the right and left bottom (bodies) are on opposite sides of a transverse axis, perpendicular to the direction of travel about which they can turn 180^0.

Reversible plough (turn wrest type)**:** A plough in which a compound body is turned from the right to the left hand by rotating it through approximately 180^0 about longitudinal axis.

Reversible process: A process that can be reversed without leaving any trace on the surroundings. Reversible processes are idealized processes, and they can be approached but never reached in reality.

Revolution: The act of revolving as the turning in a complete circuit of a body on its axis. Usually distinguished from rotation, which may mean a revolution or a part of a revolution, while the term revolution is applied to something having continuous motion, as a revolution of a shaft.

Revolutions per minute (rpm)**:** A measure of rotational speed.

Rheostat: A device for regulating electrical current, in which the current is made to flow through wires having considerable resistance. *See* Fig.6R.

Rheostatic control: A system of control, which is accomplished by varying resistance and/or reactance in the armature and/or field circuit of the driving-machine motor.

Rib: A skeleton arch, which is part of the framework that supports a vault. A small hand tool used in shaping or smoothing ware on potter's wheel. A flange extending across or around the edges of a casting, in order to strengthen a portion, which would otherwise be weak.

Rich mixture: An air-fuel mixture that has a relatively high proportion of fuel and a relatively low proportion of air.

Ridge fertilizer attachment: A series of baffles fitted to a fertilizer distributor to do fertilizer application between the ridges.

Ridge planting: Planting on ridges.

Ridge roller: Roller comprising reels, the diameter of each reel decreasing from the outer ends to center for use on ridged land so that the surface of the roller has the same form as the ridges over which they pass.

Ridger: An implement, which cut and turns the soil in two opposite directly simultaneously for forming ridge is also known as furrower. *See* Fig.7R.

Ridging: A tillage operation, which places soil into a specific configuration.

Riding plough: A plough balanced wheels and on which the operator rides during operation.

Rifling: The technique of cutting helical grooves inside a rifle barrel to impart a spinning motion to a projectile around its long axis.

Right angle: An angle of 90^0, formed by one straight line standing perpendicular to another.

Right-hand screw: A screw, which advances, when turned in clockwise rotation.

Rigid coupling: A mechanical fastening of shafts connected with the axes directly in line.

Rigid tine: A cultivator tine, which does not deflect during work.

Ring gauge: A gauge in the shape of a ring used for checking external diameters.

Ring gear: 1. The large gear on the flywheel, which is engaged, by the pinion gear when the starting motor is in action. 2. Any gear in the form of a ring, having no hub or central bore. 3. A large gear carried on the differential case within the rear-axle housing and driven by the pinion gear.

Ripsaw: An ordinary handsaw, used for sawing in the direction of the grain. Its teeth are so formed that the action is similar to that of a chisel. *See* Fig.8R.

Rittinger's law: The law that energy needed to reduce the size of a solid particle is directly proportional to the resultant increase in surface area.

Riveting: The heading over or clinching of rivets. Small rivets are headed when cold, large ones when hot.

Road band: A rim fitted to a steel wheel to prevent contact of the lugs or cleats with the road surface.

Road plates: Plates with a flat surface fitted to a track to prevent contact of the grousers with the ground.

Robot: A reprogrammable multi functional manipulator designed to move material parts, tools or specialized devices through variable programmed motions to accomplish a variety of tasks.

Rocker arm: 1. An arm pivoted at one end, the other end moving back and forth in an arc. 2. In an engine, a lever located on a fulcrum, one end on the valve stem, the other on the push rod. 3. A cam that moves with a rocking motion. *Refer* Fig.4C.

Rocket: A device where a solid or liquid fuel and an oxidizer react in the combustion chamber. The high-pressure combustion gases are then expanded in a nozzle. The gases leave the rocket at very high velocities, producing the thrust to propel the rocket.

Roller: An implement with parts mounted on horizontal shaft used for compacting the soil and crushing the clods.

Roller bearing: A bearing made of hardened-steel rollers instead of the round steel balls used in ball bearing.

Roller chain: A chain in which the links are built with cylinders or rollers to reduce noise and friction.

Roller feed fertilizer distributor: A gravity feed fertilizer distributor whose discharge mechanism consists of a rotating roller mounted transversely to the direction of travel and at the base of the hopper.

Roller feed mechanism: A feed or distributing mechanism or both consisting essentially of a fluted or plain roller at the base of a hopper.

Roller harrow: A rolling harrow with a horizontal shaft consisting of light roller with projecting teeth. *See* Fig.9R.

Roller leveling: Leveling flat stock by passing it through a machine having a series of rolls whose axes are staggered about a mean parallel path by a decreasing amount.

Roller pulverizer: A pulverizer operated by crushing action of rotating rollers.

Roller pump: A volumetric pump in which the flow of liquid is achieved by the radial displacement of rollers placed in a rotor and in contact with an eccentric stator.

Rolling: Angular motion about the longitudinal axis.

Rolling contact bearing: A bearing composed of rolling elements imposed of rolling elements interposed between an outer and inner ring.

Rolling friction: The resistance developed when a spherical or cylindrical body is rolled over a plane surface, Rollers and ball bearings are used to reduce rolling friction.

Rolling or rotary harrow: A harrow consisting of rigid or flexible tines mounted one or more horizontal or vertical shafts. It is known as gyro harrow.

Rolling resistance of non driving wheels: Force required, in the direction of the travel to overcome motion resistance of non driving wheels or transport device.

Room temperature: A temperature from 68 to 72 °F (20 to 22 °C).

Root bed: Soil profile modified by tillage or amendments for use by plant roots.

Rope drive: A system of ropes running in grooved pulleys or sheaves to transmit power.

Rot zone: The part of soil profile exploited by the roots of plants.

Rotary: A circular motion of a moving part.

Rotary atomizer: A hydraulic atomizer having the pump and nozzle combined.

Rotary blower: An incased rotating fan such as is used for forced draft in furnaces.

Rotary compressor: A positive-displacement machine in which compression of the fluid is effected directly by a rotor and without the usual piston, connecting rod, and crank mechanism of the reciprocating compressor.

Rotary cultivator: A cultivator with tines or blades mounted on a power-driven horizontal shaft (s).

Rotary duster: A duster with hand operated rotor usually carried in front or side of the operator.

Rotary engine: An engine in which the radially arranged cylinders revolve around a fixed crankshaft.

Rotary feeder: Device in which a rotating element or vane discharges powder or granules at a predetermined rate.

Rotary furnace: A heat-treating furnace of circular construction which rotates the work piece around the axis of the furnace during heat treatment; work piece are transported through the furnace along a circular path.

Rotary motion: A circular movement, such as the rotation of a crankshaft.

Rotary plough: A plough used to cut, to pulverize soil by impact forces by means of number of rotating tines or knives, which mounted, on a horizontal rotor.

Rotary puddler: A puddler with blade fixed on a horizontal shaft.

Rotary pump: A displacement pump that delivers a steady flow by the action of two members in rotational contact.

Rotary tillage: Tillage operation employing rotary action to cut, break and mix soil

Rotary valve: A valve for the admission or release of working fluid to or from an engine cylinder where the valve member is a ported piston that turns on its axis.

Rotary weeder: A ground driven implement with straight or curved spikes mounted on disc rims for destroying weeds and breaking the soil surface.

Rotary hoe: A hoe with curved pointed spikes rigidly attached to horizontal shaft.

Rotating auger plough: A plough with short plough bottom followed with vertical rotors having blades.

Rotating disc(s) distributor: A fertilizer distributor in which the centrifugal broadcasting mechanism consists of at least one rotating disc.

Rotating PTO drive shaft guard: A PTO drive shaft guard, which can rotate with the shaft except when it comes in contact with some other object.

Rotor pump: It consists of inner and outer rotors housed in the oil pump body with the rotation of pumps shaft; the inner rotor revolves with the outer rotor inner side and discharges at outer side.

Rotor: A revolving part of a machine, such as an alternator rotor, disk brake rotor, distributor rotor, etc.

Router: A two-handled tool for smoothing the face of depressed surfaces in woodwork.

Row marker: As attachment to a sow machine to mark a line on the field to guide position of extreme furrow opener during the trip.

Running-fit: The clearance between the shaft and journal to allow free running without overheating.

Runway: An artificial landing strip permitting the landing and takeoff of airplanes under all weather conditions.

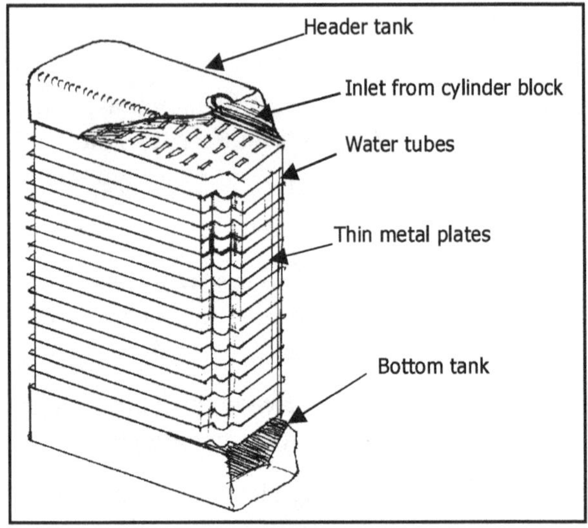

Fig. 1R Radiator

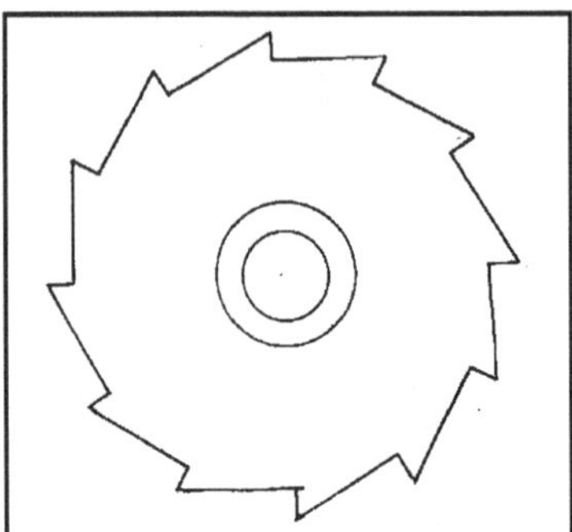

Fig. 2R Ratchet

Fig. 3R Reciprocating plate mechanism

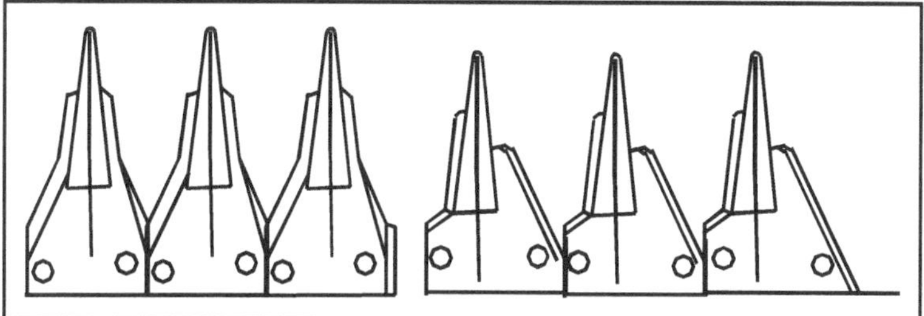

Fig. 4R Registration of cutter bar

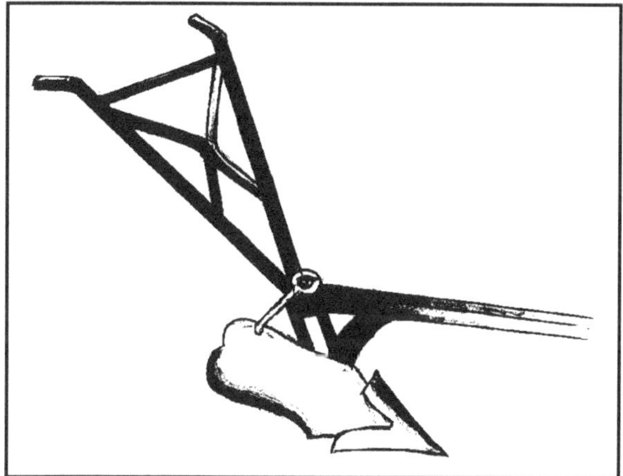

Fig. 5R Reversible plough

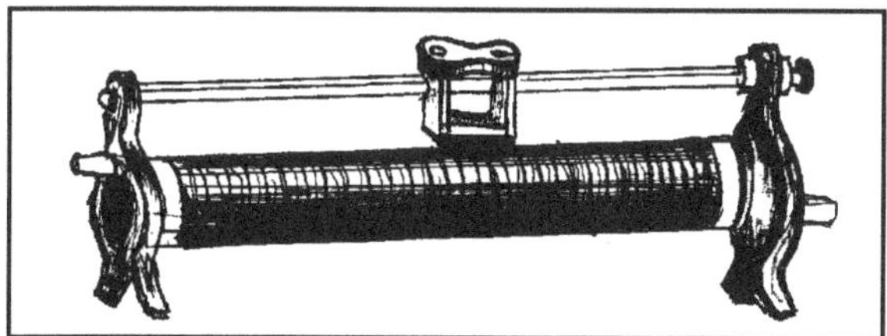

Fig. 6R Rheostat

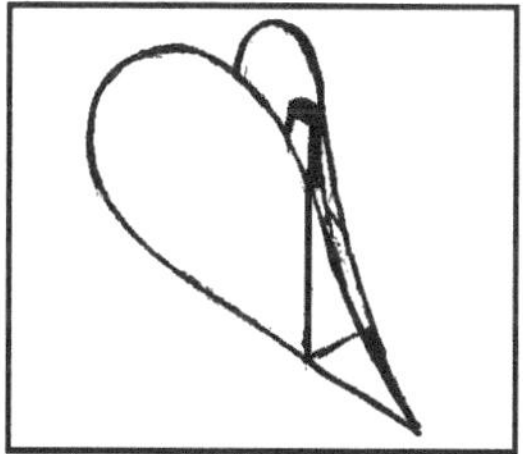

Fig. 7R Ridger

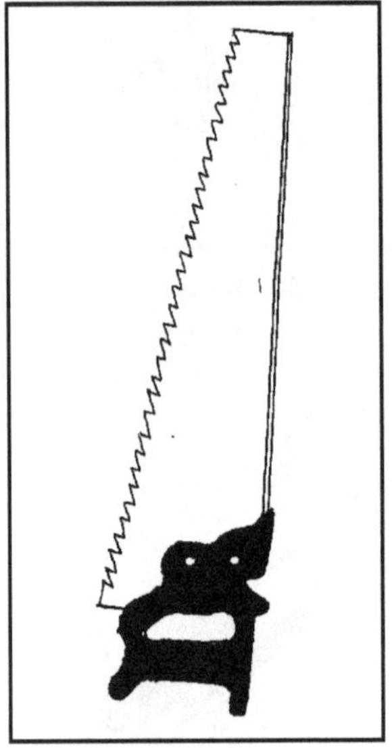

Fig. 8R Ripsaw

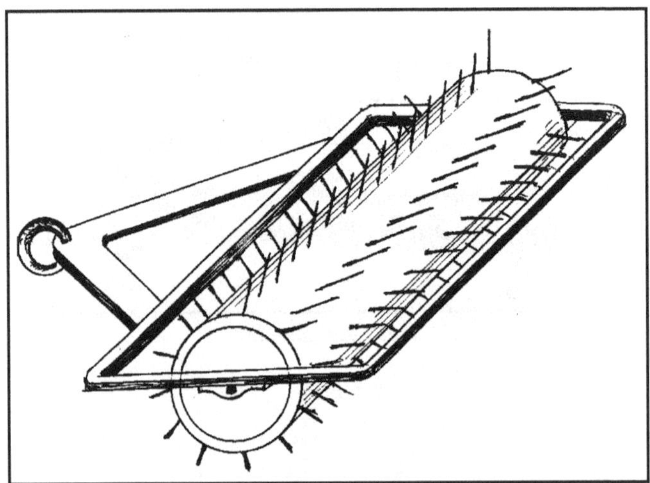

Fig. 9R Roller harrow

SAE: Society of Automotive Engineers; used to indicate a grade or weight of oil measured according to the SAE.

Safe load: The load that a piece can support without exceeding the working unit stress.

Safety device: An installation designed to protect the operator or other persons. E.g., shields, grass-catching equipment, deflectors, covers for chains, belts, drive shafts, etc.

Safety factor: The degree of surplus strength over and above normal requirements, which serves as insurance against failure.

Safety hitch: A hitch incorporating a device to avoid breakage when an implement works against an abnormal resistance.

Safety valve: 1. An automatic valve, which opens and relieves the pressure when it reaches a predetermined value, thereby protecting the system and operator. 2. A safety device for allowing steam or water to escape from a boiler when the pressure may become dangerous. Usually adjusted to permit not more than 5 pounds of pressure above maximum allowable working pressure of the boiler.

Sag: 1. To dip, bend, or cause to bend downward, a depression, especially in the middle. 2. A departure from original shape by its own weight, as the sag of a door.

Sand pump: A pump, usually a centrifugal type, capable of handling sand-and gravel-laden liquids without clogging or wearing unduly; used to extract mud and cuttings from a borehole.

Sanitation: The neutralization or removal of conditions injurious to health.

Saturated air: An air, which can hold no more moisture. Any moisture introduced into saturated air will condense.

Saturated liquid: A liquid about to vaporize.

Saturated liquid-vapour mixture: A mixture of the liquid and vapour phases that coexist in equilibrium.

Saturated steam: 1. Water vapour at the temperature of the boiling point corresponding to a given pressure. 2. Steam, which is in contact with the water from which it is generated.

Saturated vapour: A vapour that is about to condense.

Saturation (electromagnetism) : A magnetic material is said to be saturated, when, upon increasing the ampere-turns, no increase in the number of magnetic lines of force is obtained.

Saturation pressure: The pressure at which a pure substance changes phase at a given temperature.

Saturation temperature: The temperature at which a pure substance changes phase at a given pressure.

Saw blade: Metal disk with peripheral cutting teeth

Saw blade cover: Removable cover completely shielding the saw blade teeth when saw is not in use.

Saw blade guard: Device covering the rear part of the saw blade in order to protect the operator.

Saw chain: Chain serving as a cutting cool, usually consisting of drive links, cutters and guide links, held together by rivets.

Scalping: The process of cleaning in which good grains or seeds are drooped through screen openings while larger material is carried over the screen into a separate spout.

Scarification: Scratching of hard seed coat to remove the mechanical barrier to obtain uniform germination with better penetration of moisture into the seed.

Scarifying: Preparing a site for regeneration by scaring the ground surface to penetrate the covering material and expose the soil underneath.

Scavenging: 1.Removal of exhaust gas from an internal combustion engine cylinder and replacement by a fresh charge or air. 2.Removal of dissolved gases from molten metal. 3.The sweeping out, by a blast of air, of the gaseous products resulting from the firing of a gun.

Score: A scratch, ridge or groove marring a finished surface.

Scored drum: A brake drum which has become ridged or grooved; may be the result of exposed brake lining rivets, or other conditions.

Scouring: A soil tool reaction in which soil

Scrap: Pig iron that has been melted in the cupola and is to be melted again.

Scrape: To finish a surface or to fit a bearing, etc., by the use of hand tool called a scraper.

Scraper: A device to remove soil that tends to stick to the working surface of a disc.

Scraping: A method of finishing flat surfaces and bearings, etc., by the use of a hand tool called a scraper.

Screen (grain cleaning) : Perforated sheet with round of slotted holes or wire mesh used for separating brokens from whole grain.

Screening (grain cleaning): The operation of isolating the desired grains by a mechanical device where the desire grain is carried over the device and the undesired material penetrates the device.

Screws : 1. A cylindrical body with a helical groove cut into its surface. 2. A fastener with continuous ribs on a cylindrical or conical shank and a slotted, recessed, flat, or rounded head. Also known as screw fastener.

Screw adjusting caliper: Similar in construction to the firm joint caliper but has in addition a spring and screw control for securing delicate adjustments.

Screw and nut steering gear: The lower end of the steering-gear shaft has a double screw thread cut on it; one a right-hand, and the other, crossing it, a left-hand thread. When assembled, the threaded end of the shaft is engaged by two half nuts, one right-hand and one left hand. When the steering wheel is turned one half of the nut moves upward and the other half downward; this motion is transmitted to the cross shaft through rollers which operate against the lower ends of the half nuts.

Screw compressor: A rotary-element gas compressor in which compression is accomplished between two intermeshing, counter-rotating screws.

Screw conveyor: A conveyor consisting of a helical screw that rotates upon a single shaft within a stationary trough or casing, and which can move bulk material along a horizontal, inclined, or vertical plane. Also known as auger conveyor; spiral conveyor; worm conveyor.

Screw elevator: A type of screw conveyor for vertical delivery of pulverized, materials.

Screw feeder: A mechanism for handling bulk (pulverized or granulated solids) materials, in which a rotating helicoids screw moves the material forward, toward and into a process unit.

Screw processing: The process of upgrading the seeds for improving their plantability and storability.

Screw stock: Free-machining bar, rod, or wire.

Screw threads: Projections left by cutting a helical groove on a cylinder are known as screw threads. Threads may be internal or external.

Seal: A part or material that is used to close off the area of contact between two machine parts, usually to prevent fuel, gases, coolant or oil leakage.

Sealing: The act of tightly joining the cell cover to the cell compartment by means of hot pitch (sealing compound).

Seat: A surface, usually, machined, upon which another part rests or seats. e.g. the surface upon which a valved face rests.

Secateur: A scissor like tool used for pruning the plants.

Second class lever: A lever with the load applied between fulcrum and effect.

Second law of thermodynamic (Kelvin-Plank statement): It is impossible to construct an engine which while operating in a cycle produces no other effect except to extract heat from a single reservoir and to do equivalent amount of work.

Second law of thermodynamic (Clausius statement): It is impossible to construct a device which operating in a cycle, will produce no effect other than the transfer of heat from a cooler to a hotter body.

Secondary battery: The ordinary automobile lead storage battery. A secondary battery can be recharged; primary battery destroys itself as it produces electricity.

Secondary current (automobile) : The high-tension current of the automobile ignition system considered as being about 15000 volts and produced by the secondary winding of the ignition coil.

Secondary dimensions, or derived dimensions: Dimensions as velocity, energy, and volume, is expressed in terms of the primary dimensions.

Secondary shear surface: Shear surfaces, which result from twisting pushing or tumbling of the soil after or during initial displacement. Secondary shear surface are often perpendicular to the primary shear surface.

Secondary shoe: The shoe, which receives the braking action or servo action from the first shoe. The first shoe is moved around the brake drum to contact the secondary shoe and force it into contact with brake drum.

Secondary tillage: Tillage operation fallowing primary tillage, which is performed to create proper soil tilth for seeding and planting.

Secondary coil: A winding or coil, which is magnetically coupled with another coil, called a 'primary'. *Refer* Fig. 6B.

Section: A drawing representing the internal parts of an object as if it had been cut straight through vertically or horizontally. Also partial sections may be taken through particular parts.

Sectional conveyor: A belt conveyor that can be lengthened or shortened by the addition or the removal of interchangeable sections.

Sector shaft: The cross shaft of a steering gear which carries the sector roller or the teeth representing a section of a full gear.

Sediment: The matter, which settles at the bottom of a liquid.

Sedimentary rock: Rock formed under water by pressure or by cementation.

Seebeck effect: Two wires made from different metals when joined at both ends (junctions), form a closed circuit. If one of the ends is heated, as a result of the applied heat a current flows continuously in the circuit. The Seebeck effect is named in honour of Thomas Seebeck, who made this discovery in 1821.

Seed bed: The soil zone which affect germination of seeds and emergence of seedlings.

Seed broadcasting machine: A device machine used for broadcasting seeds.

Seed dresser: A machine to apply coating of protective chemicals to seeds.

Seed drill: A machine to place seeds at uniform rate at selected depth and row specified

Seed tape mechanism: A feed mechanism, which delivers a pre prepared tape of biodegradable material containing seeds at pre determined interval. Nutrients and / or crop protection chemicals may be incorporated in the tape.

Seed treating: The process of application of a thin coat of insecticide or fungicide over seeds of similar applications to improve the storability and protect viability.

Seed tube: A tube, which carries seeds from the metering device to the boot.

Seed-cum-fertilizer drill: A machine, which drills seeds and fertilizer simultaneously in the same or different rows.

Seeder: A device for placing seeds on or in the soil.

Seeding attachment: An attachment provided in a cultivator for sowing seeds.

Seed-wheel mechanism: A feed mechanism consisting of toothed wheel rotating in a horizontal plane and conveying the fertilizer through a feed gate below the star wheel.

Seize: The sticking together or adhering of surfaces, which normally move freely against each other, due to heat generated by friction caused by lack of lubrication. Such parts are often said to be 'frozen'.

Self hardening steel: Steel, which hardens when cooled in air.

Self induced current: The current produced by the self-induced e.m.f. set up in a coil when the magnetic field of the same coil changes in direction or intensity.

Self-centering: The automatic setting, locating, or marking of a cylindrical piece of work.

Self-centering chuck; A chuck, which automatically centers a piece of work through the simultaneous movement of the jaws.

Self-energizing (brake)**:** The tendency of a brake mechanism to increase the brake pressure, due to the rotation of the brake drums.

Self-inductance: The phenomena of electro-magnetic induction which takes place between the turns of a coil in a circuit.

Self-induction: The reaction of the magnetic field of a coil upon itself.

Self-propelled: Propelled and operated by a power unit or units integral with the machine.

Self-starter: An attachment for automatically starting an internal combustion engine.

Self-tapping screw: A screw designed to cut its own threads into metal into which it is run.

Semi trailer: An agricultural trailer, which, while in use, transfers a part of its total load on the towing tractor and carries the rest of the load in its axles.

Semicircular arch: An arch whose intrados is a half circle.

Semi-mounted: Partly carried on a tractor, both when in and out of operation.

Semi-portable (sprinkler system)**:** Systems designed with permanent pumping units and mains but with portable sprinkler laterals.

Sensible energy: The portion of the internal energy of a system associated with the kinetic energies of the molecules.

Separable plugs: A spark plug from which the insulator can be removed for cleaning or replacement.

Separating (grain)**:** The operation of isolating the detached grain, small debris and incompletely threshed or completely threshed grain from the bulk of straw, stem or stalk.

Separators (battery)**:** Used as insulator between the plates of a battery. It may be of wood or special composition, and must be porous to allow circulation of electrolyte.

Septic tank: A plumbing unit used for decomposing solid sewage matter. It is designed to dispose off these wastes in a completely sanitary and odorless manner by natural bacterial action, which dissolves most of the solids into liquids and gases.

Sequence: In regular order, or systematic arrangement.

Series circuit: An electric circuit in which the same current flows through all devices; positive terminals are connected to negative terminals.

Series connection: An electrical circuit wherein no current may flow through an individual part without flowing through all parts. Cells are connected in series when the positive part of one is connected to the negative of another. This connection gives the voltage equal to the voltages of all cells.

Series motor: A type of DC electric motor in which the armature and the field magnet are connected electrically in series.

Service breaking device: A device whose function is to reduce the speed of moving trailer and to bring it to halt.

Service manual: A document, which provides detailed procedure for the proper repair and/or overhaul of the equipment by a qualified mechanic.

Service pipe: The small pipe, which conveys liquids or gas from a main pipe to its places of use.

Service switch: The switch inserted at the point of entrance of the wiring to a building, controlling the whole installation in the building.

Service wires: The supply wires connecting the load in a building to the source of supply from a transformer.

Servo: A self-helping, a self-actuating device. The tendency of a brake or clutch to increase its own holding action.

Setscrew: A plain screw having a square or other shaped head, used for tightening purposes, and for locking adjustable parts in position. Setscrews are usually heat treated.

Shackle: A support by which leaf springs are attached to the car frame; the shackle permits the spring to vary in length under load. A shackle usually consists of two side plates and two bolts.

Shaft: The cylindrical machine element, which revolves in bearings and carries pulley and gears, used to transmit rotary motion and power from a driver to a driven element.

Shaft closed length: The distance between the centers of the journal cross assemblies when the PTO drive shaft is extended to the maximum length recommended by the manufacturer.

Shaft horsepower: The output power of an engine, motor, or other prime mover; or the input power to a compressor or pump.

Shaft work: The energy transmitted by a rotating shaft and is the related to the torque T applied to the shaft and the number of revolutions of the shaft per unit time.

Shank: That part of a tool by which it is connected to its handle or socket.

Shank of coulter: The part that connects the coulter to the plough beam.

Shaper: A metalworking machine on which the work is fastened to a table and the tool is moved back and fort over it by means of a sliding ram.

Share: 1. The part of the plough bottom, which penetrates into the soil and makes horizontal cut below the surface. 2. The narrow steel bar attached to the upper surface of the shoe longitudinally along the central line and projecting slightly out. *See* Fig. 1S.

Share point: The leading end of the cutting edge, which facilitates the penetration of share into the soil. *See* Fig. 1S.

Shear: The resistance of a body to being cut by the action of two parallel forces or loads acting in opposite directions.

Shear angle: The angle made by the shear plane with the work surface.

Shear blocks: The block of soil which are sheared loose from the main soil mass as a result of ultimate shearing strains applied by tillage tool.

Shear surface: Failure surfaces occurring where the soil has sheared.

Shearing die: A die with a punch for shearing the work from the stock.

Shearing machine: A machine for cutting cloth or bars, sheets, or plates of metal or other material.

Sheet metal gauge: A gauge used for measuring the thickness of sheet metal.

Shell (pod)**:** Outer hull of the pod.

Shelling: Removal of grains from cobs of kernels from pods of husk.

Shield: A part or an assembly that unless displaced or removed, restricts access to or from a hazardous area.

Shift rail: Either of the two round rod or shaft like parts, which are moved backward and forward when shifting gears. They carry the shifting forks, which, in turn, engage the gears or other parts to be moved. Shift rails have detents (depressions) in them for contact with locating pawls, which hold them in several positions to which they may be shifted.

Shifting dog: Any device used in the transmission, which is so designed as to be turned by the transmission shaft and yet may be shifted end ways to mate with other gears or part from which it receives or to which it delivers turning effort.

Shifting fork: A steel fork or yoke mounted on a sliding rod, used to move gears in and out of mesh.

Shin of mouldboard: The edge of mouldboard, which lies in a vertical plane.

Shine share: Share having shin as additional part.

Shock: The result of a sudden application of force.

Shock absorber: The hydraulic controlled device used to prevent unduly harsh spring action when the road wheels of the automobile strike obstructions on the road.

Shoe: A protective covering, as an adjustable plate, to offer sliding surface and to take up wear.

Shoe (thresher)**:** The oscillating structure which supports cleaning sieves and which may also support the chaffer extension.

Shoe (plough)**:** The wedge-shaped lower portion of the plough, which moves inside the soil.

Shoe brake: A type of brake in which friction is applied by a long shoe, extending over a large portion of the rotating drum; the shoe may be external or internal to the drum.

Shoe-type furrow opener: A furrow opener sledge shaped in elevation, with a V-Shaped leading edge and hollow body through which seeds are dropped.

Short circuit: A defect in an electric circuit that permits current to take a different path, instead of following the desired one.

Shovel: 1. A hand tool comprising a narrow blade and a wooden handle having a grip at the rear for digging, scooping and throwing the soil. 2. A soil working element with one or both end point. The later is known as reversible shovel.

Shrink fit: Where the shaft or parts are slightly larger than the hole in which it is to be inserted. The outer part is heated above its normal operating temperature or the inner part chilled below its normal operating temperature and assembled in this condition; upon cooling an exceptionally tight fit is obtained.

Shrinkage: The failure of a casting to retain its exact size, weight, and shape during the process of cooling in the mold. Shrinkage is much grater in large castings than in small.

Shrinking: Drawing together; contracting; diminishing. The contraction of a casting in cooling.

Shunt: A parallel connection or circuit.

Shunt for ammeter: A predetermined known resistance connected in parallel with the galvanometer; used to limit the current through the meter. An ammeter is basically a galvanometer connected in parallel with a known resistance.

Shunt machine: The electrical generator with the armature winding and field winding connected in parallel so that the current flowing through one does not flow through the other.

Shunt motor: A type of DC electric motor that connects the magnet and armature in parallel. Used when the motor speed must be constant, irrespective of variation in load.

Shut-off nozzle: A nozzle with shut off device, which may be used without changing parts.

Shutter (air flow control) **:** A part of an appliance to control the flow of air.

SI system: The metric system of measurement.

Sickle: A curved steel blade having a handgrip used for harvesting by manual power.

Sickle bar mower: A cutting machine, which utilizes a power source to operate a knife or knives to provide a shearing action with stationary cutter bar or movable knife.

Side cap: A plate fitted above a landside.

Side delivery reaper: A reaper, which delivers the harvested crop on the side having the space for the next run.

Side draft: The horizontal component of the pull perpendicular to the direction of motion. This is developed if the centre of resistance is not directly behind the centre of pull.

Sieve aspirator (rice grader)**:** A rice grader which separates brokens from rice mass by means of vibrating screen in conjunction with air stream. The traces of bran adhering to head rice and brokens are separated by an aspirator.

Sieve separator: A simple separator employing an oscillating sieve.

Sieving: The operation of separating the desired grains by a mechanical device where the desired grains penetrate the device and the undesired material is carried over the device,

Silicone: A plastic family based on silica with a wide range of physical forms, from simple fluids and greases to flexible rubbers and hard, durable resins. Since they offer excellent weather, heat, and wear properties, they are used in gaskets, circuit breakers, and aircraft parts where high temperature operation is required. They are also used in lubricants, and adhesives.

Silicon copper: A rich copper alloy added to molten copper in order to secure clean, solid castings free form blowholes, swellings, etc.

Silicon steel: Steel containing from 1 to 2 per cent of silicon is used in the manufacture of springs. Steel containing 3 to 5 per cent possesses the magnetic qualities, which make it valuable for electromagnets.

Silver: A white, ductile, malleable metal atomic number 47; an excellent conductor of heat and electricity. Melting point, 961^{0}F; specific gravity 10 to 11according to purity. Symbol Ag

Simple cooling: The process of lowering the temperature of atmospheric air when no moisture is removed.

Simple engine: An engine (such as a steam engine) in which expansion occurs in a single phase, after which the working fluid is exhausted.

Simple heating: The process of raising the temperature of atmospheric air when no moisture is added.

Simple machine: Any of several elementary machines, one or more being incorporated in every mechanical machine; usually, only the lever, wheel and axle, pulley (or block and tackle), inclined plane, and screw are included, although the gear drive and hydraulic press may also be considered as simple machines.

Single acting: A machine, or device, in which the motive power is applied in one direction only; action is effective in one direction only.

Single action disc harrow: A disc harrow with two gangs placed end-to-end, displacing the soil outwardly. *Refer* Fig.10D.

Single aspirating separator: A separate with as single aspirating leg.

Single belting: A belt formed of only a single thickness of leather.

Single plate clutch: A popular clutch consisting of three plates, usually two driving plates (flywheel and pressure plate) and one driven plate (clutch plate). So called because of the single driven plate. *Refer* Fig. 15C.

Single wheel plough: A swing plough provided with a wheel or toe at the front.

Single-disc furrow opener: A furrow opener consisting of one concave disc.

Skeleton wheel: A steel wheel to give a minimum ground contact area.

Skew gear: Gears with spiral teeth. Skew bevel gears are now largely used on the rear axles of automobiles; they increase strength and promote smooth and quiet action.

Skid ring: A circumferential flange on the rim of a front wheel.

Skidder (tree) **:** Self propelled machine designed to transport trees or parts of trees by dragging.

Skidding (tree) **:** Transporting trees or parts of trees by dragging.

Slab: 1. A thin piece of stone, marble, concrete, or the like, having a flat surface. 2. The outside pieces cut from a log when sawing it into boards. 3. A thick rectangular piece of steel for rolling into plates.

Slack: Looseness or play in a mechanism.

Slade: A horizontal plate bearing on a furrow bottom, usually integral with a landside.

Slasher (tree) **:** Machine designed to cross cut felled trees to pre-determined lengths.

Slat mouldboard: A mouldboard the surface of which is made of slats placed along the length of the mouldboard, so that there are gaps between the slats. It is often used in sticky soils. *See* Fig.2S.

Sledge: A long handled heavy hammer used with both hands.

Sleeve: A hollow tube or cylinder, which surrounds a rod or shaft.

Sleeve bearing: A machine bearing in which the shaft turns and is lubricated by a sleeve.

Sleeve nut: A threaded member made of steel or cast iron fastened to the cone through cone head, which can be made to move forwards or backwards on the sleeve for adjusting the position of the cone.

Slice or slice bar: A firing tool, used for breaking up and separating the clinkers on a furnace grate.

Slicing cut: A cut taken with the sliding or slicing motion for the purpose of removing thin pieces.

Slide calliper: A pocket calliper consisting of a graduated bar which slides in a retaining piece.

Slide slip: The sidewise motion of a rolling tyre, which is not properly aligned.

Sliding gear: A gear fitted to a shaft in such a way that it can not turn on a shaft, but can be made to slide along the shaft to engage or disengage from other gears.

Sling psychrometer: A device with both a dry-bulb thermometer and a wet-bulb temperature mounted on the frame of the device so that when it is swung through the air both the wet-and dry-bulb temperatures can be read simultaneously.

Slip: Relative movement in the direction of travel at the mutual contact surface of the traction device and the surface, which supports it.

Slip nose share: A share with detachable point.

Slip share: A one-piece share with curved cutting edge, having no additional part.

Slipping clutch: The clutch, which because of wear, or some other reason is no longer able to hold the full engine power and therefore permits the flywheel and pressure plate to revolve faster than the driven plate is turned.

Slit nozzle: A hydraulic spray nozzle with an opening in the shape of a slit producing a flat sheet of spray.

Slot: A long, narrow groove, particularly one cut to receive some corresponding part of mechanism.

Slotter: A machine tool used for making a mortise or shaping the sides of an aperture.

Sludge: A composition of oxidized petroleum product along with an emulsion formed by the mixture of oil and water. This forms a pasty substance and clogs oil lines and passages and interferes with engine lubrication.

Smoke box: Chamber external to a boiler for trapping the unburned products of combustion.

Snagging: Removing surplus metal or large surface defects by using the grinding wheel.

Sod mouldboard (breaker mouldboard): *Refe*r breaker mouldboard.

Soil abrasion: The scratching, cutting or abrading of tillage tools caused by the action of soil.

Soil additives: Foreign material, other than seeds, which are added to or incorporated in soil for directly influencing the soil condition or environment. These include pesticides fertilizer mulches, or conditioners but not foreign bodies; such as drain tiles, which have an indirect influence.

Soil adhesion: The sticking of soil to tillage tools or wheels.

Soil compaction: Reduction in the specified volume of soil by means of mechanical manipulation.

Soil cone: An adhered soil body, which resembles a cone.

Soil cultivation: Tillage operation performed to create soil conditions conductive to improve aeration infiltration, compaction and moister conservation as well as to control weeds.

Soil cutting: Soil separation from soil mass by slicing action.

Soil failure: The alteration or destruction of soil structure caused by mechanical forces, such as shear, compression and impact.

Soil heaving: The swelling of soil resulting from natural forces such as freezing

Soil injector: Appliance for injecting chemicals into soil.

Soil pipe: Term generally applied to cast iron pipe in 32.5 cm lengths used for house drainage.

Soil reaction: The reaction caused by the application of mechanical forces to the soil.

Soil scoop: An implement used to collect and move the soil from one spot to another. *See* Fig.3S.

Soil shatter: The general fragmentation of a soil mass resulting from the action of the tillage forces.

Soil sheet: An adhered soil body, which covers a large area of a tool like sheet.

Soil slicing: The slicing of soil across a surface.

Soil surgeon: An implement consisting of number of rows of 'U' shaped knives fixed to the underside of the rectangular board used for breaking clods and uprooting weeds.

Soil wedge: An adhered soil body, which resembles a narrow wedge.

Soil working surface: Portion of tool, implement or machine designed to be in contact with soil in order to effect proper manipulation.

Soldering: An uniting under the proper heat, of pieces of metal by means of the dissimilar metal or alloy.

Sole plate: A foundation plate to which a piece of machinery is bolted.

Solenoid break: A device that retards or arrests a rotational motion by means of the magnetic resistance of a solenoid.

Solenoid switch: A switch, which is opened and closed electro magnetically by the movement of a solenoid core. The core also causes a mechanical action, such as the movement of a drive pinion into mesh with flywheel teeth for cranking.

Solenoid: An electromechanical device (a coil of wire wound around a movable core) that, when connected to an electrical source such as a battery, produces a mechanical movement. This movement can be used to control a valve or to produce other movements.

Solenoid valve: Valve actuated by a solenoid, for controlling the flow of gases or liquids in pipes.

Solid box: A solid, an unadjustable ring bearing lined with babbit metal, used on light machinery.

Solid cone nozzle: A Cone nozzle in which extra liquid enters in the swirl chamber centrally from its base so that the air core is filled to from a solid cone of the droplets.

Solid coupling: A flanged face or a compression type coupling used to connect two shafts to make a permanent joint capacity of a shaft; a solid coupling has no flexibility.

Solid friction: The friction which results when the surface of one solid body is move across the surface of another solid body.

Solid injection system: A fuel injection system for a diesel engine in which pump forces fuel through a fuel line and an atomizing nozzle into the combustion chamber.

Solid mouldboard: A mouldboard with unbroken surface.

Solid phase: Molecules arranged in repeated three-dimensional pattern (lattice). Because of the small distances between molecules in a solid, the attractive forces of molecules on each other are large and keep the molecules at fixed positions.

Solid Stream spray: Spray with a cylindrical shape.

Solubility: It represents the maximum amount of solid that can be dissolved in a liquid at a specified temperature.

Sorting: 1. Collecting similar items. 2. It involves inspection, followed by separation according to variety, size, colour, degree of ripeness, maturity, defects, etc. it is actually a separating operation supplementary to grading.

Source(s) of ignition: Devices or equipment that, because of their modes of use or operation, are capable of providing sufficient thermal energy to ignite flammable compressed natural gas-air mixtures when introduced into such a mixture or when such a mixture comes into contact with them and that will permit propagation of flame away from them.

Sowing: The process of placing of seeds, in seedbed.

Spacers: Mild steel strips ranging from 0.25 to 0.63 mm in thickness inserted between unchamfered cage bars to provide gap.

Spacing (precision) drill: A seed drill in which the metering mechanism distributes seeds singly or in groups at a set spacing in continuous parallel rows.

Spacing-drill with perforated belt: A spacing drill in which the metering mechanism consists of a perforated belt. The spacing are adjusted by varying the belt speed or by selecting belts with differently spaced perforations

Spade (powrah): A hand tool comprising a broad blade and a wooden handle used for loosening, cutting, earthing, dressing and removing the soil.

Spade lug: A wedge-shaped metal projection fitted to the rim of a steel wheel.

Spandrel: The irregular triangular space between an arch and the beam above the same.

Spanner: A type of wrench used for tightening or loosening up of nuts.

Spark: A discharge of electrical energy; used to produce combustion of fuel gas in an engine.

Spark arrester: A device fitted with the exhaust system of an engine to arrest the glowing carbon particles.

Spark plug electrode. The metal conductor through the center of the insulator of the spark plug. Also, the points between which the spark is caused to jump.

Spark plug: A device that screws into the cylinder head of an engine provides a spark to ignite the compressed air-fuel mixture in the combustion chamber. It consists of a threaded outer shell with an outside electrode, insulator, and a copper gasket. The spark gap is in between 0.50 to 0.85 mm. Two types of spark plug are 1. Cold plug and 2. Hot plug. *See* Fig.4S.

Spark plug (cold plug): It has a short insulator, extending into the cylinder. Due to a short path conducts heat rapidly, dissipates to engine cylinder jacket and get cooled.

Spark plug (cold plug): It has comparatively larger insulator. Heat is not dissipated quickly as it has to cover longer path.

Spark ignition system: In the automobile, the system that furnishes high-voltage sparks to the engine cylinders to fire the compressed air-fuel mixture. There are two methods in spark ignition as 1. Battery ignition and 2. Magneto ignition. Refer battery ignition system and magneto ignition system.

Spark-ignition (SI) engine: The reciprocating engine in which the combustion of the air-fuel mixture is initiated by a spark plug.

Sparking at brushes: Small sparks or flashes between commutator and brush due to poor contact, improper position of brush, irregular surface, dust on commutator etc.

Specific fuel consumption: The mass of fuel consumed per unit of work. For example kg/Kw.hr.

Specific gravity, or relative density: The ratio of the density of a substance to the density of some standard substance at a specified temperature (usually water at 4°C, for which the density is 1000 kg/m^3). Or ratio of the weight of the gas to the weight of air. Or the ratio of the weight of a liquid to an equal volume of pure water.

Specific ground pressure: The quotient of front axle mass (or combine mass in case of full-track-laying combines) and ground contact area giving a rating comparable to tyre pressure in case of wheeled combine.

Specific heat: The relative amount of heat required to raise the temperature of a unit mass of a substance by one degree compared with the amount of heat required to raise the temperature of the same weight of water by one degree.

Specific heat at constant pressure (Cp): The energy required to raise the temperature of the unit mass of a substance by one degree, as the pressure is maintained constant. It can be defined as the change in the enthalpy of a substance per unit change in temperature at constant pressure.

Specific heat at constant volume (Cv): The energy required to raise the temperature of the unit mass of a substance by one degree as the volume is maintained constant. It can also be defined as the change in the internal energy of a substance per unit change in temperature at constant volume.

Specific volume: The reciprocal of density and is defined as the volume per unit mass.

Specific weight: The weight of a unit volume of a substance and is determined from the product of the local acceleration of gravity and the substance density.

Specifications: Information provided by the manufacturer, describing systems and their components, operations, and clearances. Also, the service procedures that must be followed for a system or component to operate properly.

Spectrophotometer: An instrument for forming and studying spectra. Spectra produced by vaporized substances are useful in the study of their composition.

Sped cone: A cone shaped pulley, or a pulley composed of a series of pulleys of increasing diameter forming a stepped cone.

Sped lathe: A metalworker's lathe, not equipped with mechanical feed.

Speedometer: An instrument for recording distance traveled and the rate of speed in kilometers per hour. Usually mounted kin such a position as to be clearly visible by the car operator.

Speedometer drive gear: The small gear on the transmission shaft, which turns the speedometer driven gear from which the flexible shaft of the speedometer receives its turning effort.

Spike tooth cylinder type thresher: A thresher, the threshing unit of which consists of drums having rows of spikes with a closed cylinder casing and concave, and equipped with a set of sieves and an aspiratory blower.

Spike-tooth harrow: A harrow with pegs as working part fitted on rigid, articulated or flexible frame. *See* Fig.5S.

Spiral: A curve formed by a fixed point moving about a center, and continually increasing or decreasing the distance from it.

Spirit level: An instrument for testing the horizontal and vertical accuracy of work. It consists of a glass tube or bulb nearly full of spirit and enclosed in a wood or metal case. When the bubble in the tube is in a central position, it indicates the accuracy of the work.

Splash lubrication system: Troughs are provided at the lower end of connecting rod, which while rotating splashes oil out of the pan. The splashing action of oil develops a fog or mist of oil that lubricates inner parts of engine such as bearings, cylinder walls, piston, piston pins, timing gears etc. This system is suitable for single cylinder engine having closed crankcase.

Spline: The ridged and grooved arrangement on the end of the shaft made to fit into mating ridges and grooves within a hallow part so as to permit the hallow part to move endways over the spline, but cause it to turn with the shaft.

Split pulley: A pulley made in halves and bolted together.

Spoke: One of the arms of a wheel connecting the hub with the rim.

Spool (spacer): The flanged tube mounted on the gang axle between every two discs to retain them at fixed position on the shaft.

Spot welding: Method of electrical resistance welding in which overlapping edges are united at intervals by passing a heavy current through two copper electrodes which are at the same time forced together by mechanical, hydraulic or pneumatic power.

Spray angle: The angle formed close to a hydraulic spray nozzle by the edge of the spray.

Spray area: Area covered by the spraying expressed in hectare. In case of overhead spraying, this will be equal to the land area; in case of strip spraying, this will depend on row width and strip length; and in case of plantation spraying, the height of the tree and strip length.

Spray boom: A device on which the nozzles are mounted and which may form or support one or more pipelines which are carrying the liquid to the nozzle.

Spray dryer: A machine for drying an atomized mist by direct contact with hot gases.

Spray gun: An apparatus shaped like a gun, which delivers an atomized mist of liquid.

Spray lance boom: Spray pipe attached to the end of spray lance on to which a number of nozzles are fitted.

Spray: The droplets produced by a nozzle.

Sprayer: An appliance used for spraying

Spraying: The division and emission into the air of a liquid or a spray mixture in the form of droplets.

Spray lance: A hand-held tube through which the liquid, after being released from cutoff device, reaches to nozzle.

Spray leg (drop leg): An auxiliary vertical spray boom fixed below a main horizontal spray boom.

Spray mixture: Liquid containing the diluted formulated product ready for spraying.

Spray nozzles: A part or an assembly of parts having orifice, which transforms the fluid being ejected under pressure into a spray.

Spray pond: A pond where warm water is sprayed into the air and is cooled by the air as it falls into the pond.

Spray rate (pesticide)**:** Amount of active ingredient of pesticide applied per unit area.

Spray tank: The part of the sprayer, which contains the spray liquid or mixture.

Spray volume: Total volume of spray mixture applied to an area.

Spreading Rotor: A rotating auger, disc, or series of toothed or paddle shaped blades at the rear of a manure spreader, which spreads the manure.

Spring rate: The distance a spring is compressed under a certain load. E.g., a 200 of load will cause a spring to be compressed twice as much as a 100 load.

Spring tine: A cultivator tine made of spring steel, which can deflect when an obstacle is encountered in the soil.

Spring tooth harrow: A harrow with tough and flexible teeth designed primarily to work with hard and stony soils. *See* Fig.6S.

Spring work: is the work done to change the length of a spring.

Spring-loaded tine: A rigid tine hinged to the frame and loaded with one or more spring to enable it to swing on encountering an obstacle.

Sprocket: A tooth like projection on a wheel shaped so as to engage with the links of a chain.

Sprocket chain: A continuous chain, which meshes with the teeth of a sprocket and thus can transmit mechanical power from one sprocket to another.

Spray nozzle: A device in which a liquid is subdivided to form a stream of a small drops.

Spur gear: Gears provided with teeth running straight across the gear face.

Stacking: Pilling of timber in specified sizes.

Standard: The part connecting the soil-cutting unit of implement to the beam, in a curved beam, the standard is the part of the beam.

Standard conditions: Temperature and pressure conditions that correspond to 0 °C (32 °F) and 14.7 psi (1 bar) respectively. Sometimes, standard temperature is taken as room temperature.

Star drill: A tool with a star-shaped point used for drilling in stone or masonry.

Star wheel mechanism: A feed mechanism consisting of a series of toothed wheels rotating in a horizontal plane in the base of hopper, each conveying fertilizer under a feed gate.

Stare wheel distributor: A gravity feed fertilizer distributor in which the fertilizers is discharged by means of notched discs (star wheels) through adjustable slots.

Starling engine: A type of internal combustion engine in which the piston is moved by changes in the pressure of a working gas that is alternately heated and cooled. It has two isothermal processes and two constant: volume processes.

Starter (starting motor): A series electric motor used on automobiles to engine until it starts under its own power.

Starting crank: A crank formerly used for turning over the engine in order to start it. Since the invention of self-starters, the starting crank is rarely used except during certain types of repair jobs.

Starting switch: The electrical device used to close the starting motor circuit and thus cause the starting motor to operate.

Starting torque (electrical motor): The turning effort produced by a motor upon its shaft through the electromagnetic effect at the initial flow of current.

Static load: A nonvarying load; the basal pressure exerted by the weight of a mass at rest, such as the load imposed on a drill bit by the weight of the drill-stem equipment or the pressure exerted on the rocks around an underground opening by the weight of the superimposed rocks. Also known as dead load.

Static pressures: The pressure exerted by a static fluid medium on the wall of the container.

Stationary (knife) coulter: A stationary blade fixed downward in a vertical position on a beam.

Stationary engine: An engine located on a fixed foundation, as distinguished from a portable engine.

Stationary thresher: The thresher, which performs the threshing operation while the machine itself being stationary.

Stator: The fixed part in an a.c. motor or generator, that part which does not rotate.

Steam engine: A thermodynamic device for the conversion of heat in steam into work, generally in the form of a positive displacement piston and cylinder mechanism.

Steam generator: The combination of a boiler and a heat exchanger section (the superheater), where steam is superheated.

Steam jacket: A casing applied to the cylinders and heads of a steam engine, or other space, to keep the surfaces hot and dry.

Steam power plant: An external-combustion engine in which steam (water) is the working fluid. That is, combustion takes place outside the engine, and the thermal energy released during this process is transferred to the steam as heat. A turbine in the power plant converts some of the energy of the steam into rotating shaft work.

Steam superheater: A boiler component in which sensible heat is added to the steam after it has been evaporated from the liquid phase.

Steam: The gas into which water is converted by boiling.

Steel pulley: Usually of the split type made of sheet steel. Light in weight and satisfactory in operation.

Steel wheel: A wheel made of steel generally fitted with spade lugs or cleats.

Steerage: Allowing limited lateral movement under manual control independent of the tractor.

Steering arm: An arm that transmits turning motion from the steering wheel of an automotive vehicle to the drag link. *See* Fig.7S.

Steering arm: The arm attached to the steering knuckle and used to turn it. Also the third arm on the left hand steering knuckle. *See* Fig.7S.

Steering brake: Means of turning, stopping, or holding a tracked vehicle by braking the tracks individually.

Steering gear ratio: The ratio of the worm and gear within the steering gear.

Steering knuckle arms: Two irregular-shaped forgings, one left and one right, attached to the steering knuckles in order to make connection with the steering mechanism.

Steering knuckle: A part of the front axle or front wheel suspension, which is turned in order to move or steer the front wheel.

Steering knuckle spindle: That part of the steering knuckle carrying the wheel hub on its bearings.

Steering system: The system, governing the angular movement of front wheels of a tractor is called steering system. This system minimizes the efforts of the operator in turning the front wheel with the application of leaverages.

Steering relay: An idler arm, similar to pitman arm, carrying the steering gear rod used to transmit transverse motion from the pitman arm in certain designs of steering linkage.

Step block or step bearing: A bearing, which takes the end thrust of vertical shafts. Also known as pivot bearing.

Stepdown transformer: A transformer which changes the supply or line voltage to a lower value.

Stepped grain bed: An oscillating bed or pan on which the grain and chaff mixture fall from the cylinder and straw walker.

Stiffener: Any angle, plate, channel, or other shape riveted to a member to increase rigidly.

Stirling engine: An engine in which work is performed by the expansion of a gas at high temperature; heat for the expansion is supplied through the wall of the piston.

Stoichiometric air or theoretical air: The minimum amount of air needed for the complete combustion of a fuel. When a fuel is completely burned with theoretical air, no uncombined oxygen will be present in the product gases.

Stoichiometric air-fuel ratio: The exact air-fuel ratio required to completely react a fuel into water and carbon dioxide.

Stoichiometric combustion or theoretical combustion: is the ideal combustion process during which a fuel is burned completely with theoretical air.

Stone trap: A device mounted on the feeding side of the cylinder to trap the stones or any other similar objects in order to avoid any damage to the crop conveying and threshing units.

Stop bearing: A device, which supports the bottom end of a vertical shaft. Also known as pivot bearing.

Stopwatch: A watch, which can be started or stopped by Pressing a lamb used in timing races,heating scientific readings in particular time interval.

Storage battery: The lead and acid electrical unit, known as secondary battery, used to receive the charging current from the generator, where current causes certain chemical changes in the battery. When these chemical changes take place in opposite direction, the battery becomes a source of electrical supply.

STP: Standard Temperature and Pressure.

Straight share: A share with straight cutting edge.

Strainer (Sprayer): A component for separating foreign matter from the spray liquid.

Strakes: An assembly of metal lugs or cleats attached to a pneumatic tyred wheel.

Straw rack: An oscillating rack assembly of one-piece construction, which separates the remaining grains from straw.

Straw rack area: The product of the straight length and the inside width of the separator side structure immediately adjacent to the straw walker or rack expressed in square millimeters.

Straw raddle area: The product of the raddle length and exposed width of the raddle expressed in square millimeters.

Straw spreader-cum-cutter: A component used to cut the straw in pieces coming out from walker or rack and to spread in the field.

Straw walker: The assembly of two or more racks, which agitates the straw and separates the remaining grains from straw.

Stress analysis: Investigation of the distribution of stresses within a material combined where possible with their measurement. In qualitative analysis only the distribution of stresses is ascertained, in quantitative analysis individual stresses are measured. Techniques are electronic diffraction, electronic strain gauge, photo elasticity, etc.

Stress: An internal force, which resists change in the shape or size of a body.

Strip spraying: The spraying performed on the field between the rows of the crop.

Strip tillage: A tillage system in which only isolated band of soil are tilted.

Stripper: A device used on presses to prevent the punched metal from lifting with the punch.

Stroke: The distance moved by the piston from top lead center to bottom dead center in one direction.

Stroke bore ratio: The ratio of the distance travelled by a piston in a cylinder to the diameter of the cylinder.

Stub runner: A furrow opener consisting of two tapered flat pieces of steel welded together so as to form a narrow V-shaped leading end and a hollow body into which seeds are dropped.

Stubble mouldboard: A short but broader mouldboard with relativity abrupt curvature, which lifts, breaks and turns the furrow slice used in suitable stubble soils. *See* Fig.8S.

Stubble plough: A plough specially designed to turn over and bury stubble.

Studded cylinder distributor: A gravity feed fertilizer distributor in which the fertilizers are discharged from a hopper by cylinder fitted with studs.

Studded-roller drill: A seed drill, the metering mechanism of which consists of studded rollers. The seed rate is controlled by the Varying speed of the rotating shaft.

Studded- roller feed mechanism: A feed mechanism with series of rollers having regular peripheral projections, which meter seeds from a seed hopper. *See* Fig.9S.

Stuffing box: A recess chamber, through which piston rods or valve stem passed, surrounded there in with packing, and is used to prevent leakage.

Stump-jump plough: A plough with spring loaded plough bottom or discs which rides over stumps and other obstruction in the soil.

Sub irrigation: Application of water by percolating from wide spaced furrows to an impermeable layer in the soil with a temporary rise of water table to wet the entire soil body. This practice requires an impervious sub-soil layered, a porous surface soil and a uniform topography of moderate slope.

Subsoiling: Chiseling at depth greater then 40cm.

Sublimation: The process of passing from the solid phase directly into the vapour phase.

Sublimation line: A line that separates the solid and vapour regions on the phase diagram.

Sub-soiler: A plough designed to penetrate the soil to depths more than those achieved during normal ploughing operation, usually 40 cm or more.

Suction: Suction exists in a vessel when the pressure is lower than the atmospheric pressure;

Suction lift: The head, in feet that a pump must provide on the inlet side to raise the liquid from the supply well to the level of the pump. Also known as suction head.

Suction strainer: A device situated at the bottom of the tank or on the end of the suction preventing undesirable solid from entering the spray system.

Suction strainer: A device situated at the bottom of the tank or on the end of the tank, which prevents undesirable solids from entering the tank.

Suction stroke: The intake stroke on which the fuel mixture is drawn into the cylinder. Refer Fig. 9F.

Sugarcane crusher: A machine used for extracting juice from sugarcane.

Sugarcane planter: A machine used for planting sugarcane sets in the field.

Sugarcane stripper: A device used for removal of dry leaves and leave sheaths from cane.

Sugerbeet harvester: A machine, which lifts, tops and/or cleans the tubers and delivers them ready for carting.

Sugerbeet lifter: An implement to loosen and lift sugar beets from the soil.

Sulky plough: A single bottom riding plough.

Sulphated: Plate Battery plates which have become hardened because they have absorbed and held more than normal amounts of Sulphuric acid from the electrolyte of the battery solution.

Sun gear: The gear at the center of the any planetary gear set about which the planetary gears rotate.

Super charger: A mechanical device for supplying the engine with a grater weight of charge than would normally be induced at the prevailing atmospheric pressure and temperature. There are several types, viz., centrifugal, positive drivers, rotary blower, and turbo. A mechnaicl device designed to give positive fuel mixture to racing motors and airplane motors. The mixture is put into the motors much more rapidly than is done in stock carburetion.

Supercharging: A method of introducing air for combustion into the cylinder into the cylinder of an internal combustion engine at a pressure in excess of that which can be obtained by natural aspiration.

Supplemental irrigation: Irrigation, which is used primarily to supplement rainfall.

Surface grinding: The operation of grinding plane or flat metal surfaces.

Surface irrigation: Irrigation where the soil surface is used as a conduit as in furrow and border irrigation.

Surface tension: is the force per unit length used to overcome the microscopic forces between molecules at the liquid-air interfaces.

Surface water: Water stored on or flowing over the surface of the soil.

Swath: Cut material as left by a machine.

Swath board: An attachment to an outer shoe to direct the swath away from the uncut crop.

Swathing attachment: An attachment to mower cutter bar to concentrate the swath.

Sweat: To coat with solder the surfaces to be joined, then causing the surfaces to adhere by the application of heat, as opposed to the common method of making a joint by direct application of the soldering iron.

Sweating: When parts to be soldered are tinned, and then heated sufficiently to melt the solder without the use of the soldering iron, the process is called " sweating".

Sweep: A two winged cutting point, which cuts the soil in a horizontal direction under the surface.

Swing over handle: A handle that pivots from one end of a walk behind mover to the other end to allow reversing the direction of travel of the mower without turning the mower around.

Swing plough: An animal-drawn plough without supporting wheel whose stability is ensured by the handles, the depth and width of work being regulated by using a draft adjustment.

Swirl back plate: Part of a particular type of turbulence nozzle, which forms the rear part of the swirl chamber and the tangential liquid entry channels.

Swirl chamber: The cavity between the nozzle disc and the swirl plate.

Swirl plate: The part of cone nozzle, which imparts rotation to the liquid passing through it. It is also known as swirl core.

Switch: A device that opens and closes an electric circuit.

Synchromesh transmission: A transmission designed to synchronize, i.e. equalize the speeds of mating gears before they are meshed.

Synchronize: To cause to agree in time; happen simultaneously.

Synchronized: To cause two events to occur at the same time. E.g. time a mechanism so that two or more sparks will occur at the same instant.

Synchronized shifting: Changing speed gears, with the gears being brought to the same speed before the change can be made.

Synchronizer drum: The part of the synchronizing mechanism of the transmission which carries the internal synchronizing cone, and which contacts the external cone to equalize the speeds of meshing parts before meshing is completed.

Synchronizing cone: The cone like parts of the transmission, which are engaged by the synchronizing member so as to equalize speed when shifting gears in order to prevent clashing.

Synchronizing spark timing: Opening of two sets of contact points within a distributor so as to called two sparks to occur at the same instant or to cause alternate sparks to occur in correct degree travel. Also setting two sets of contact points opening the same circuit to open at the exact instant.

Syndicator type thresher: A thresher, the threshing unit of which consists of a corrugated flywheel with serrated chopping knives and a closed cylinder casing and concave. This is also known as chaff cutter type thresher.

Synthetic: Chemical compound made from elements or simpler compounds, especially a substance that duplicates a natural one.

—□—□—

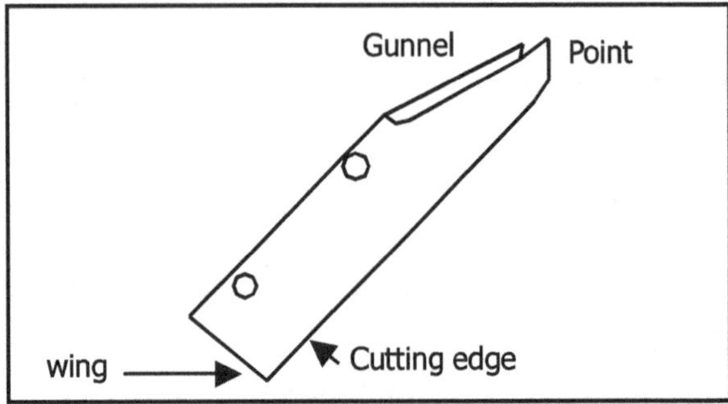

Fig. 1S Share

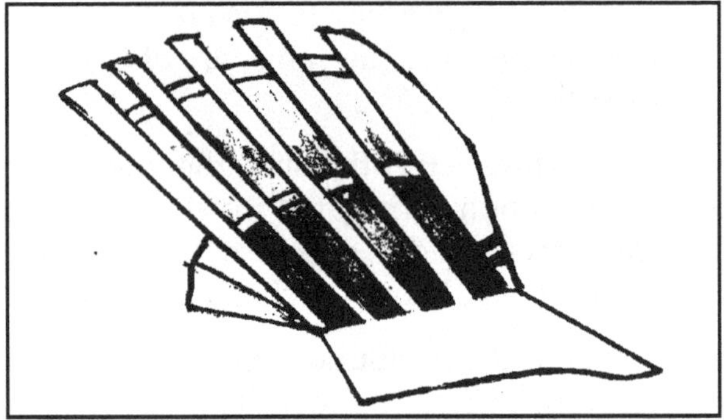

Fig. 2S Slat Mouldboard

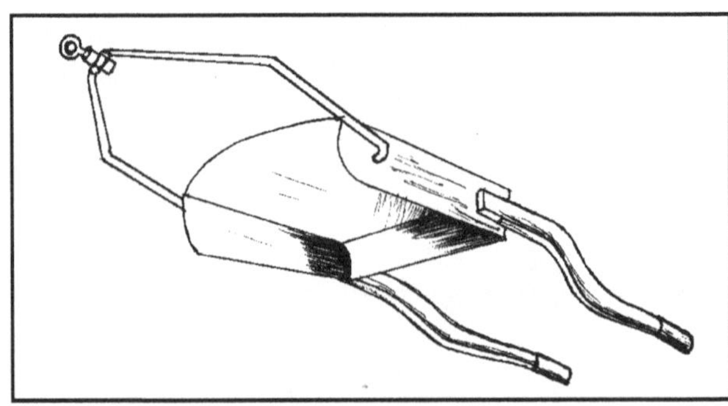

Fig. 3S Soil scoop

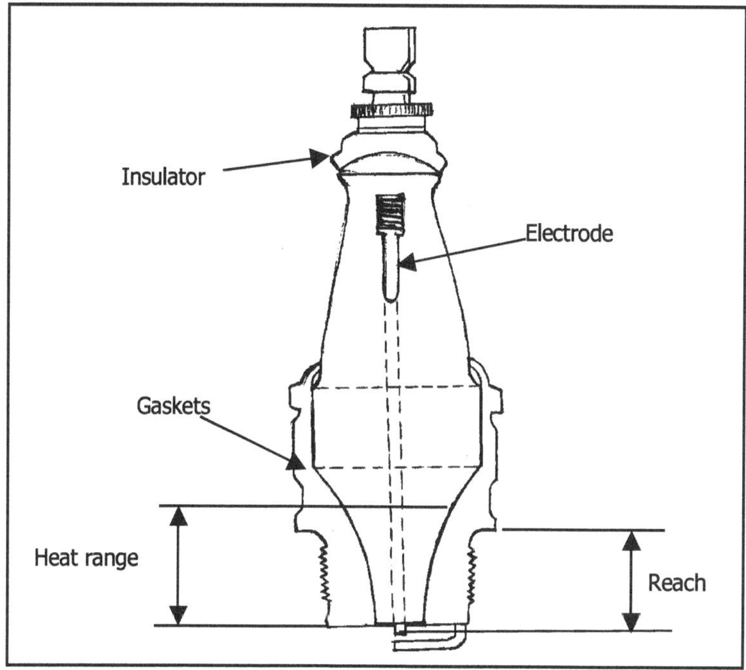

Fig. 4S Spark plug

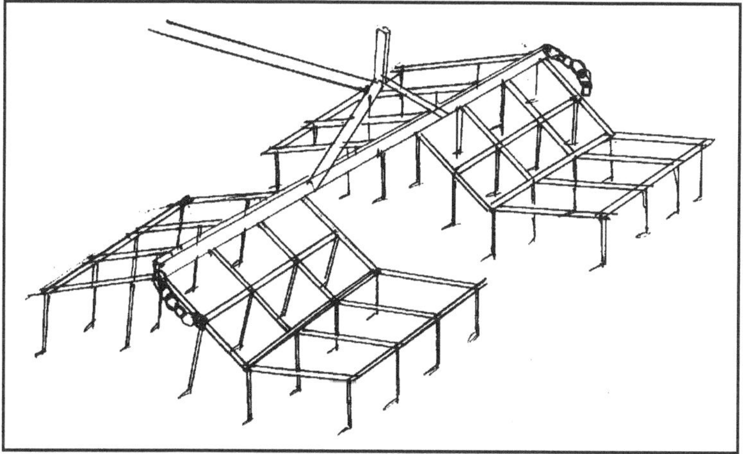

Fig. 5S Spike tooth harrow

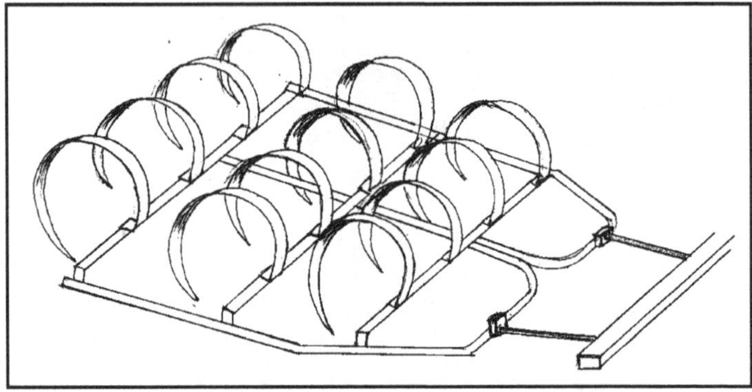

Fig. 6S Spring tooth harrow

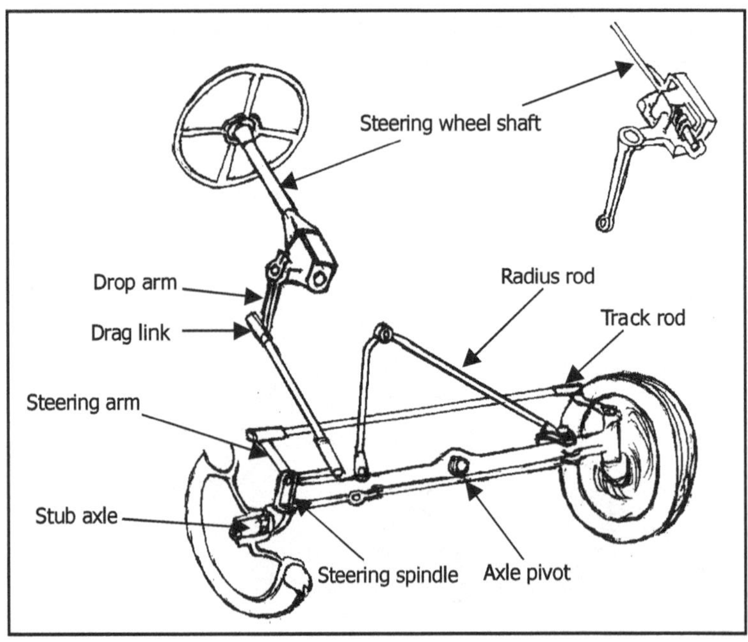

Fig. 7S Steering system

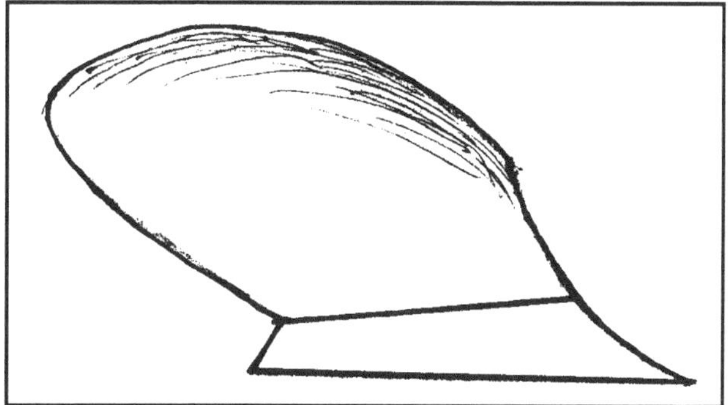

Fig. 8S Stubble Mouldboard

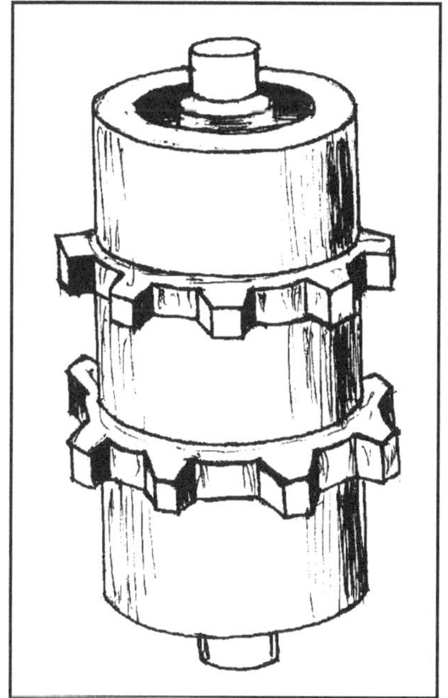

Fig. 9S Studded roller feed mechanism

T bolt: A bolt shaped like the letter 'T', the head being a transverse piece, which fits into the recessed under cut T slots, as on a table of a milling machine or planning machine.

T head engine: A cross section of an engine block resembles the letter T. Valves are arranged on both sides of the engine; requiring two camshafts and two camshafts drive gears. An expensive type of construction.

Tachometer: An instrument for measuring the speeds of shafts in revolutions per minute.

Tail pulley: A pulley at the tail of the belt conveyor opposite the normal discharge end; may be drive pulley or an idler pulley.

Tail screw: The screw, which operates the spindle of the tailstock on a lathe.

Taillight: A signal light, required by law to be carried on the rear of every automobile, and lighted when the car is driven at night.

Tailstock: The movable head of a lathe as distinguished from the headstock, which is fixed.

Take up: A tensioning device in a belt conveyor system for taking up slack of loose parts.

Take up pulley: An adjustable idler pulley to accommodate changes in the length of the conveyor belt to maintain proper belt tension.

Tandem disc harrow: It is a disc harrow comprising of four gangs in which each gang can be angled in opposite direction. See Fig. 1T.

Tangent: Touching a line or surface at point without intersecting.

Tangential velocity: The instantaneous linear velocity of a body moving in a circular path, its direction is tangential to a circular path at the point in question.

Tank (fuel) **:** The container for fuel on an automobile. *Refer* Fig.11F and 12F.

Tap bolt: A bolt usually threaded for its entire length. It is finished only on the point and the underside of the head. Tap bolts are made with both square and hexagon heads.

Tap: 1. The process of cutting threads with a tap. 2. A fluted, threaded tool for cutting female or inside threads.

Taper: A gradual and uniform decrease in size, as a tapered socket, a tapered shift, a tapered shank.

Taper gauge: A gauge for testing the accuracy of tapers, either inside or outside.

Taper pin: Made of round stock, used for fastening some part to a shaft. It is graded in size by no. from 1 to 10. Number 1 is 0.156-inch diameter at the large end, and is from $3/4^{th}$ to 1 inch long. Number 10 is 0.706-inch diameter at the large end, and 1 ½ to 6 inch long.

Taper reamer: Reamer of the ordinary fluted type for reaming tapered holes; as a reamer used to prepare a hole for a tapered pin.

Tappet: 1. A reciprocating part between the cam and the valve. 2. Adjusting device for varying the clearance between the valve stem and the cam. May be built in the valve lifter in an engine or may be installed in the rocker arm on an overhead valve engine. *See* Fig.2T.

Tappet rod: A rod carrying a tappet or tappets, as one for opening or closing the valves in a stem or an internal combustion engine. *See* Fig.2T.

Tapping: 1. Forming an internal screw thread in a hole or other part by means of a tap. 2. Opening the pouring hole of a melting furnace to remove a molten metal.

Teak: An East Indian tree of large size. The wood is very durable, and is highly prized for ship building and for furniture.

Tear strength: The force needed to initiate or to continue tearing a sheet or fabric.

Technology: The branch of knowledge that deals with application of science and basic engineering principles.

Tee: A lifting for connecting pipes of unequal sizes, or for changing direction of pipe runs.

Template: Any temporary pattern, guide, or model, by which work is either marked out, or by which its accuracy is checked.

Tenacity: That property by which material resists forces tending to tear it apart.

Tensile: Pertaining to extension or tension, as tensile action.

Tensile strain: A strain or pull in a longitudinal direction; a reverse of a crushing strain.

Tensile strength: The strength necessary to enable a bar or structure to resist a tensile strain.

Tensile stress: The stress to which a bar or structure is subjected when in tension.

Tension pulley: A pulley around which an endless rope passes mounted on a trolley or other movable bearing so that a slack of the rope can be readily taken up by the pull of the weights.

Terminal: The end of an electrical cable, the small metallic part attached to an electrical wire or cable, the point to which an electric wire or cable is attached.

Testing: A procedure for determining whether mechanical device, prime mover or machinery is in proper workable condition or not.

Testing set: Those instruments or devices used for determining whether wiring or equipment is in perfect working order.

Thermal: Have or pertaining to heat.

Thermal conductivity: The ability of a metal to transmit heat through its mass.

Thermal efficiency of a heat engine: The ratio of the network produced (W_{ne}) by a heat engine to the total heat input (Q_{in}.) ; $h_{th} = W_{net}/Q_{in}$.

Thermal efficiency of an engine: The ratio of the energy output of an engine to the energy in the fuel required to produce that output.

Thermal energy reservoir: A large body of infinite heat capacity, which is capable of absorbing or rejecting an unlimited quantity of heat without suffering appreciable change in its thermodynamic co ordinates.

Thermal equilibrium: Means that the temperature is the same throughout the entire system.

Thermal resistance: A measure of a body's ability to prevent heat from flowing through it. It is equal to the difference between the temperatures of opposite faces of the body divided by the weight of heat flow.

Thermal shock: Stress produced in a body or in a material as a result of undergoing a sudden change in temperature.

Thermal valve: A valve controlled by an element made of material that exhibits a significant change in properties in response to a change in temperature.

Thermocouple: The electrical generator made by the welding together, at one end, of two dissimilar metal rods or wires, which produce electricity at the free ends when the weld is heated; used in pyrometers for measuring extreme heat.

Thermodynamics: 1. Science, which deals with the energies possessed by the gases and vapuors. 2.Can be defined as the science of energy. The name thermodynamics stems from the Greek words *therme* (heat) and *dynamis* (power), which is most descriptive of the early efforts to convert heat into power. Today the same name is broadly interpreted to include all aspects of energy and energy transformations, including power production, refrigeration, and relationships among the properties of matter.

Thermodynamic cycle or cyclic process: When a process or processes are performed on a system in such a way that the final state is identical with the initial state, it is then known as thermodynamic cycle or cyclic process.

Thermodynamic process: When a system changes its state from one equilibrium state to another equilibrium state, then the path of successive state through which system has passed is known as thermodynamic process.

Thermodynamic system: 1. Defined as a quantity of matter or a region in space chosen for study. Also known as 'system'. 2. A definite area or space, where attention is focused for studying thermodynamic process. Thermodynamic system has a boundry and any thing outside the boundary is surrounding.

Thermopile: A group of dissimilar material arranged alternately in such a manner that the junctions can be heated to produce an electric current.

Thermosiphon system: This type of cooling system is based on the fact that hot water rises and cold water settles in the system. Water heated by the engine rises within it and flows to the top of the radiator where it is cooled; by the time it reaches the bottom of the radiator it is relatively cool and starts on its round again.

Thermostat: A heat-controlled valve is used in the cooling system of an engine to regulate the flow of water between the cylinder block and the radiator.

Third arm: The second arm on the left hand steering knuckle body used to receive the forward end of the drag link when fore and aft steering is used.

Third law of thermodynamics: States that the entropy of a pure crystalline substance at absolute zero temperature is zero.

Thread : 1. Helical ridges machined on the surface of the circular shaft. The purposes of threads are: (a) to hold parts together, i.e., screws, bolts, nuts, etc.; (b) to tighten parts together to withstand the pressure of liquids or gases, i.e., pipes, fittings, etc.; (c) to transmit power as through jack screws, worm drivers, etc.; (d) to adjust with great accuracy the parts of such instruments as calipers, micrometers, etc. 2. A fine cord or filament, usually of non-metallic material, as silk, cotton, or wool. Also see screw thread.

Thread cutter: A tool used to cut screw threads on a pipe, screw, or bolt.

Thread gauge: A gauge for checking a pitch of screw threads.

Threading: The cutting of screw threads, either internal or external.

Threading machine: A tool used to cut or form threads inside or outside of cylinder or cone.

Threads per inch: Thread size. It is a standard practice to fix the no. of threads per inch depending on diameter, as for ½ inch diameter, 13 threads per inch; for 1 inch diameter, 8 threads per inch, etc.

Thread point linkage: *See* Fig.3T, 4T & 5T.

Throat clearance (MB plough): It is the perpendicular distance between point of share and lower position of the beam of plough. See Fig. 6T.

Throttle: To shut off or regulate.

Throttle valve: 1. A thin, flat disk valve placed in a pipe or opening for the purpose of closing it partly or entirely, as in controlling the flow of air to be mixed with fuel in an automobile engine. 2. A valve in a steam line for controlling the flow. *Refer* Fig.1A.

Throttle system (governor): In this system, the explosion intensity is changed by controlling the amount of charge entering during the suction stroke. The number of explosions is not varied, only explosion intensity is changed. Throttle system of governor is divided into three types as 1. Centrifugal governor, 2. Pneumatic governor and 3.Hydraulic governor. See centrifugal governor, pneumatic governor and hydraulic governor.

Throw piston : The amount of eccentricity, as in the crankshaft of an engine. The throw is equal to half the length of the stroke of the piston.

Thrust bearing: A bearing, which sustains axial loads and prevents axial movement of a loaded shaft.

Thrust load: A load or pressure parallel to or in the direction of the shaft of a vehicle.

Tie: A piece inserted or attached to other pieces to hold them in position.

Tie rod: The transverse rod connecting the front wheels of an automobile in order to permits used in tying up composed types. Or the portion of the front axle assembly used to connect the radius arm of the right and left hand steering knuckle bodies.

Tie rod ball: The ball bolt so made as to permit universal motion within predetermined limits of the tie rod or drag link. Usually used on the end of the tie rod or drag link.

Tight fit: A fit made with light pressure.

Tight pulley: A pulley attached to its shaft; as opposed to a loose pulley, which runs free on its shaft.

Timer ignition : A device used on automobiles to break primary ignition circuit at the proper time for the spark to occur in cylinders.

Timing belt pulley: A pulley that is similar to an uncrowned flat-belt pulley, except the grooves for the belt's teeth are cut in the pulley's face parallel to the axis.

Timing gears: Any group of gears which are driven from the engine crankshaft to cause the valves, ignition and other engine driven accessories to operate at the desired time during the engine cycle. Since the camshaft has to turn once for every two revolutions of the crankshaft, these gears must be in the ratio of 2 to 1.

Timing mark: Usually marks placed on the flywheel or dynamic balances to indicate top dead centre and the no. of degrees before top dead centre at which ignition should occur.

Tip radius: The distance of the outermost point of a propeller blade from the axis of rotation. Also called as propeller radius.

Title block: The outlined space usually in the lower right corner, or in strip form across the bottom of a drawing, containing name of company, title of drawing, scale, date, and such other information as may be considered necessary.

Toe in: The setting of the front wheels so that they are closer together in front than in the rear.

Toe-out: The outward inclination of the wheels of an automobile at the front on turns due to setting the steering arms at an angle.

Toggle switch: An electrical switch which is operated by a knob or a projecting arm moving it up and down or to side.

Tolerance: Allowable inexactness or error in the dimensions of manufactured machine parts. Also called limit or allowance.

Tool maker: A workman skilled in the making of jigs, fixtures, gauges, etc.

Tool room: A room in which tools and parts are stored and from which they are issued to workmen; a shop where jigs, fixtures, etc., are made, stored, and repaired.

Top dead center (TDC) : The position of the piston when it forms the smallest volume in the cylinder. *Refer* Fig. 2A.

Topping cycle: A power cycle operating at high average temperatures that rejects heat to a power cycle operating at lower average temperatures.

Torque: The ability of a force to move a load or to do work. Turning or twisting motion. It is measured in standard in pound-feet (ft-lb), or in metric in the Newton-meter (Nm).

Torque wrench: A wrench that indicates the amount of twisting effort (rotary motion) being applied with the wrench.

Torricelli (torr): Unit of pressure equal to 1 mm of mercury (1 mm Hg) or 1 atm = 760 torr. = 760 hg. Named for the Italian physicist *Evangelista Torricelli* (1608:1647), who first measured the pressure of the atmosphere using a tube of mercury upended within a dish of mercury, coincidentally producing the first vacuum within the tube.

Torsion: The act of twisting. The tendency to deform, as a rod, by twisting.

Total energy of a flowing fluid: The sum of the enthalpy, kinetic, and potential energies of the flowing fluid.

Toughness: The resistance of a metal to begin permanent deformation and the further resistance to failure after permanent deformation has begun.

Traction: The condition of friction as between the tyre and surface of the roadway; the holding of the tyre on the roadway; the ability to transmit power from the tyre to the roadway.

Tractor: 1. An automotive vehicle having four wheels or caterpillar trade used for pulling agricultural or construction implements. 2. The front pulling section of semi-trailer. Also known as truck tractor.

Trammel: A beam compass, in which head slides along a straight bar. It is tightened by setscrews, and is used to strike radii to large for the capacity of an ordinary compass.

Transducer: Any device that converts an input signal of one form into an output signal of a different form.

Transfer machine: 1. Equipment that moves parts from one production location in a factory to another. 2. A device that holds a work piece and moves it automatically through the stages of a manufacturing process.

Transformer: An instrument by means of which electrical current are changed in regard to voltage and amperage from high to low or vice versa.

Transistor: A compact unit performing many of the functions of vacuum tubes in electronic circuits with the advantage of small size, an active semiconductor device with three or more electrodes.

Transmission: The gear box or the housing containing speed changing gears used on the automobile for securing varying ratios of drive, second and third speeds, and also reversing. Power transmitted from the engine is received in the gearbox and transmitted at varying speeds to the propeller or drive shaft and then to the gear axle.

Transmission shaft: The main shaft of transmission usually shaft carrying the first and second gears, which are driven from the gear cluster on the counter shaft.

Transmitter: 1. In telephony, that part of the instrument into which one speaks. It consists of two flat carbon electrodes, one mounted on vibrating diaphragm, and the other stationary. Between them is a mass of granular carbons.

Transparent: Transmitting light so that objects may be distinctly seen.

Triangulation: The use of series of triangles formed by lines which connect points of observation, for measuring distances and areas of land and water.

Triple line: The locus of the conditions where all three phases of a pure substance coexist in equilibrium. The states on the triple line of a substance have the same pressure and temperature but different specific volumes.

Triple point of water: The state at which all three phases of water coexist in equilibrium.

Trouble diagnosis (trouble shooting) : A process of diagnosing or locating the source of trouble from observation and testing.

Try square: A small square used by mechanics in testing squareness of their work; also used to lay off right angles.

T-square: A draftsman's tool, consisting of a blade from 2 to 3 inch wide and from 1 to 5 feet long, attached at right angles to a head, which is at least twice as thick as a blade; used for ruling parallel, horizontal lines.

Tuner: A coil condenser circuit, which can be adjusted to select a desired radio signal and reject others.

Tune up: A process of accurate and careful adjustment to obtain the best engine performance.

Tungsten: A hard, brittle, white or grayish metal, a rare element of the chrome group, used in contact points and lamp filaments.

Tungsten lamp: A type of incandescent lamp having a filament of fine tungsten wire.

Turbidity: The degree of cloudiness of water as compared to perfectly clear water; due to silt, etc.

Turbine: 1. A series of angled blades located on a wheel against which fluids or gases are impelled to rotate a shaft. 2. a device that produces shaft work due to a decrease of enthalpy, kinetic, and potential energies of a flowing fluid. 3. A type of steam engine in which all driving parts rotate.

Turbocharger: A supercharger driven by the engine exhaust gas pressure for pressurizing the intake air or air-fuel charge of an engine, so as to increase the mixture delivered to the cylinders thus increasing power output.

Turbofan (or fan-jet) engine: is the most widely used engine in aircraft propulsion. In this engine a large fan driven by the turbine forces a considerable amount of air through a duct surrounding the engine. A turbofan engine is based on the principle that for the same power, a large volume of slower-moving air will produce more thrust than a small volume of fast-moving air. The first commercial turbofan engine was successfully tested in 1955.

Turbulence: 1. A disturbed, irregular motion of the fluids or gases. 2. A swirling motion given incoming fuel charges by the shape of the cylinder head. This condition is desirable as a means of reducing knocking and pinging.

Turning machine: A machine used to prepare the edge of a cylindrical or flaring body, Inside/outside threading etc.

Turret lathe: A lathe with revolving tool head, making possible several operations without removing the tools from the machine.

Two-phase alternator: An alternator which delivers two sets of current differing in phase displacement by 90 deg.

Two phase: Two windings or circuits with a phase displacement of 90^0 (electrical).

Two-way valve: A mechanical device that controls the flow of fluid by allowing flow in either of two directions.

Two-cycle engine: An engine design permitting a power stroke once for each revolution of the crankshaft.

Two-stroke cycle: An internal combustion engine cycle completed in two strokes of the piston.

— — —

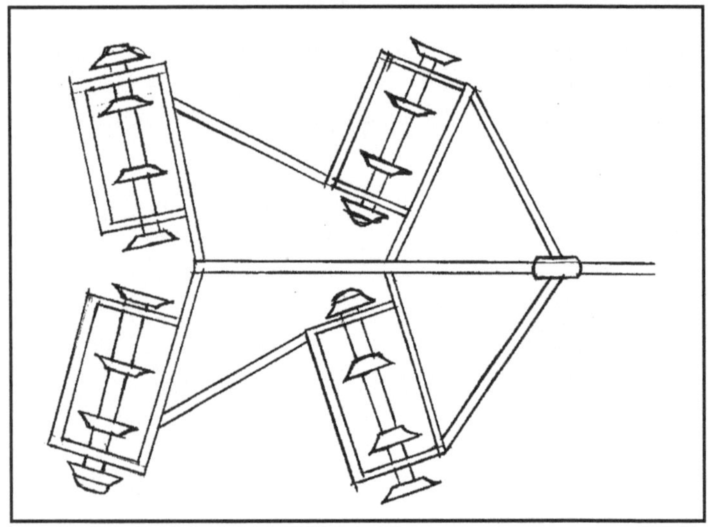

Fig. 1T Tandem disc harrow

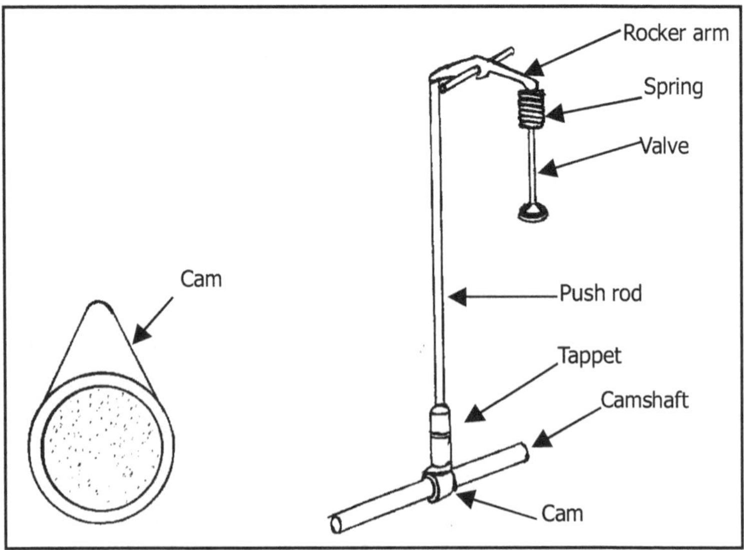

Fig. 2T Position of Tappet and Rocker arm

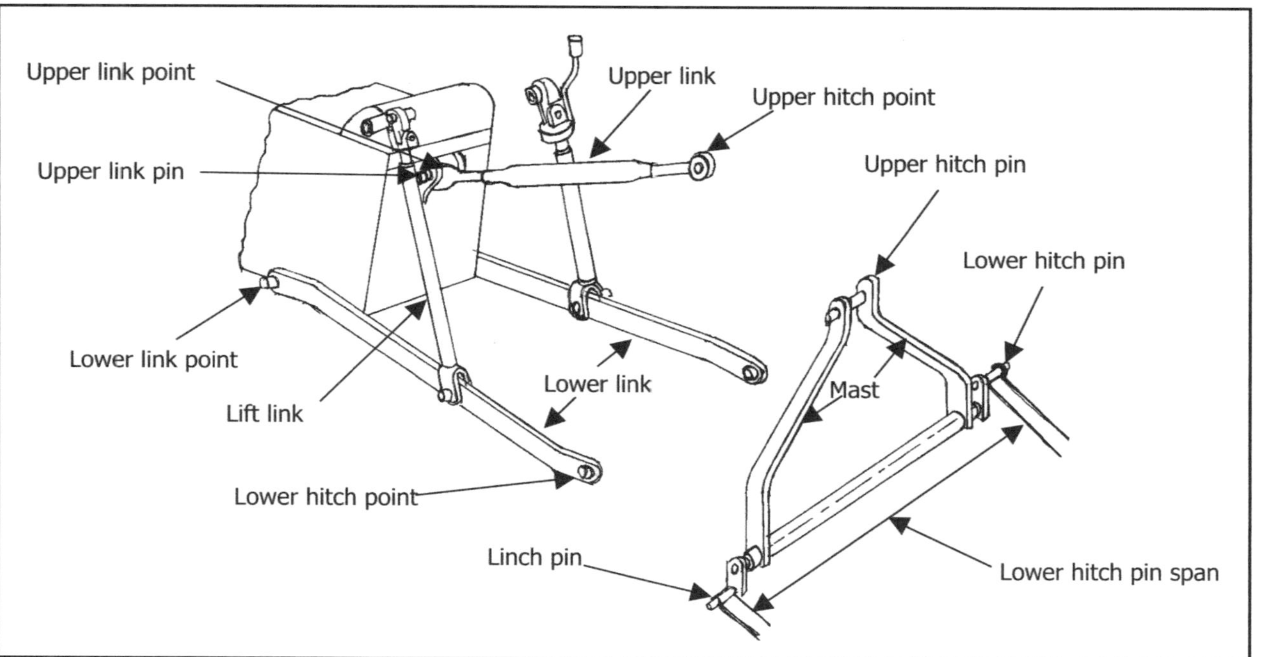

Fig. 3T Three point linkage

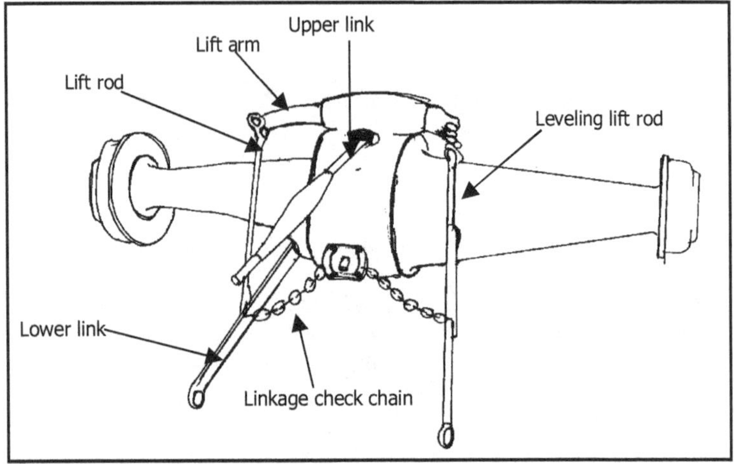

Fig. 4T Three point linkage check chain

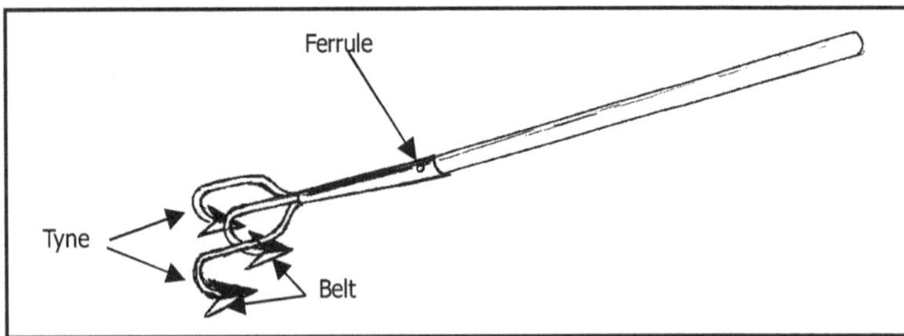

Fig. 5T Three tyne gubbler

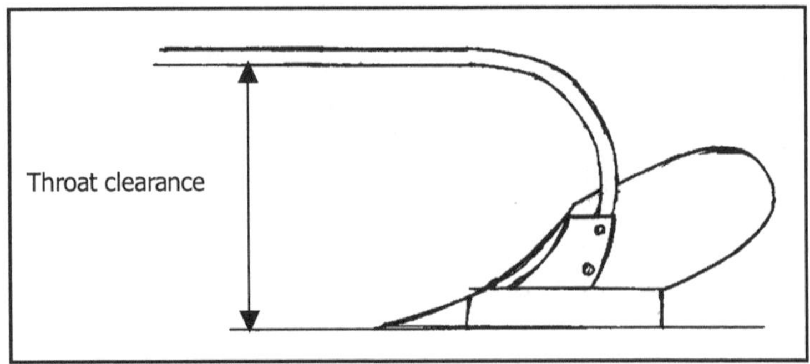

Fig. 6T Throat clearance

U clamp: A clamp shaped like the letter 'U' used for clamping down work on planer beds.

Ultimate strength: The highest unit stress that can be sustained, this occurring just at or just before rupture in stress strain diagram.

Ultrasonic drilling: A vibration drilling method in which ultrasonic vibrations are generated by the compression and extension of a core of electrostatic or magneto static material in a rapidly alternating electric or magnetic field.

Ultrasonic machining: The removal of material by abrasive bombardment and crushing in which a flat-ended tool of soft alloy steel is made to vibrate at a frequency of about 20,000 Hz and an amplitude of 0.001-0.003 inch (0.0254-0.0762 millimeter) while a fine abrasive of silicon carbide, aluminum oxide, or boron carbide is carried by a liquid between tool and work.

Uniform: Implies no change with location over a specified region.

Union: A coupling or connection for pipes.

Unit of illumination: Candle power- The brightness of a lamp. Mean spherical candlepower is candlepower averaged over all directions from the center of the lamp. Mean horizontal candlepower is average candlepower in a horizontal plane through the luminous center of a lamp.

Unit of magnetic flux: The total number of lines of force set up in a magnetic substance. Treated as a magnetic current flowing in a magnetic circuit.

Unit of magnetic intensity: Magneto-motive force (M.M.F.), Magnetic pressure, which drives lines of force through a magnetic circuit.

Unit stress: The stress on a unit of section area, usually expressed in kg per square meter.

Universal gas constant (Ru): It is the product of the gas constant and the molecular mass of the gas, Ru is the same for all substances and its value is 8.314 kJ/kg mol ° K.

Universal grinding machine: A grinding machine provided with a swivel table, a swivel headstock, and a swivel wheel head, used for external and internal cylindrical grinding, surface grinding, face grinding, etc.

Universal joint: A type of coupling which permits the free rotation of two shafts whose axes are not in a straight line.

Universal saw table: A saw table that may be tilted to permit sawing at a bevel.

Universal testing machine: A machine used for testing the strength and elasticity of materials.

Unsteady flow: Fluid flow in which the volumetric relation of two or more phases liquid-gas, liquid-liquid and so on vary along the cause of flow, can be the result of changes in temperature, pressure, or composition. It is also called as transient flow.

Updraft carburetor: For a gasoline engine a fuel-air mixing device in which both the fuel jet and the airflow are upward.

Updraft furnace: A furnace in which volumes of air are supplied from below the fuel bed or supply.

Upper explosive limit (UEL): The UEL is the highest gas concentration that will support an explosion when mixed with air, contained and ignited. The UEL is always lower than the UFL.

Upper flammable limit (UFL): The UFL of a gas is the highest gas concentration that will support a self-propagating flame when mixed with air and ignited. Above the UFL, there is not enough oxygen present to support combustion. The air-fuel mixture is too rich.

Uranium: A hard malleable metal used to increase strength and toughness of high-speed steels. Also used in small quantities in cast iron. Two principal isotopes are U-235 and U-238. Only U-235 is readily fissionable and only 1 part in 140 of natural uranium is U-235.

— — —

V belt: An endless power transmission belt with a trapezoidal cross section, which runs in a pulley with a V-shaped groove. The top surface of the belt approximately flush with to top of pulley. V belt transmits higher torque at less width and tension than a flat belt.

V blocks: Metal blocks cut 'V' shaped on one side to permit holding cylindrical work while machining or testing.

V type engine: An engine with cylinder blocks arranged in a 'V' shape. One crankshaft serves both banks of cylinders.

Vacuum: Negative gauge pressure, or a pressure less than atmospheric pressure. Vacuum can be measured in psi, but is usually measured in inches or millimeters of mercury (Hg); a reading of 30 in. Hg (762 mm Hg) would indicate a perfect vacuum.

Vacuum brake: The mechanical breaking system operated by a vacuum taken from the intake manifold or carburetor at a point just above the throttle valve.

Vacuum cleaner: A motor-driven fan machine used for sucking up dirt and dust from rugs, etc.

Vacuum control: The term may apply to any automobile devices such as brakes, clutch, etc., controlled by manifold vacuum.

Vacuum cooling: A way to cool a substance by reducing the pressure of the sealed cooling chamber to the saturation pressure at the desired low temperature and evaporating some water from the products to be cooled. The heat of vaporization during evaporation is absorbed from the products, which lowers the product temperature.

Vacuum freezing: The application of vacuum cooling when the pressure (actually, the vapour pressure) in the vacuum chamber is dropped below 0.6 kPa, the saturation pressure of water at 0°C.

Vacuum gauge: 1. A device that measures pressures below atmospheric. 2. An instrument calibrated in atmospheric inches to test the vacuum produced in intake manifold of an engine or in the fuel line.

Vacuum pressure: The pressure below atmospheric pressure and is measured by a vacuum gage that indicates the difference between the atmospheric pressure and the absolute pressure.

Vacuum pump: A compressor for exhausting air and no condensable gases from a space that is to be maintained at sub atmospheric pressure.

Vacuum spark advance: A metal diaphragm connected by a rod to the breaker-point plate. The diaphragm is actuated by the intake manifold vacuum, which varies with the throttle opening, causing an advance of the spark by rotation of the breaker-point plate.

Vacuum tube: An electron tube evacuated to such a degree that the electrical characteristics are essentially unaffected by the presence of residual gas or vapour.

Valve: 1. A device, which control the flow of intake and exhaust gases to and from the engine cylinder. 2. A device that can be opened or closed to allow, to regulate or to stop the flow of a liquid or gas. *See* Fig.4C.

Valve clearance: The air gap allowed between the end of the valve stem and the valve lifter or rocker arm to compensate for expansion due to heat.

Valve face: The beveled part of the valve head, which mates with the valve seat.

Valve follower: A linkage between the cam and the push rod of a valve for maintenance of alignment.

Valve grinding: Also called valve lapping. A process of lapping or mating the valve seat and valve face usually performed with the aid of an abrasive.

Valve head: The disk part of a poppet valve that gives a tight closure on the valve seat *or* The portion of valve, upon which the valve face is machined.

Valve lap: The amount of lap between the opening of the intake and the exhaust, indicated by marks on the flywheel in degrees or number of teeth. Opening of the intake valve prior to closing of the exhaust valve is called 'positive lap'. When the intake valve opens after the closing of the exhaust valve it is called "minus lap".

Valve lifter: A push rod of plunger placed between the cam and the valve of an engine; is often adjustable to vary the length of the unit.

Valve lifter assembly: The various parts, guides, tappets, adjusting screws, etc., which operate between the camshaft and the poppet valves to regulate valve action. These parts are sometimes mounted in a housing bolted to the crankcase.

Valve margin: On a poppet valve, the space or rim between the surface of the head and the surface of the valve face.

Valve retainer: It may also apply to any locking device, which keeps the valve and spring in position.

Valve seat: The matched surface upon which the valve face rests.

Valve seat inserts: Rings of alloy steel pressed in position in the valve ports to increase the life of the valve seats.

Valve spring: A spring attached to a valve to return it into the seat after it has been released from the lifting or opening means. *Refer* Fig.4C.

Valve stem: The rod by means of which the disk or plug is moved to open or close a valve. It rests within a guide. *Refer* Fig.4C.

Valve stem clearance: Distance between lower end of valve stem and valve tappet. Its purpose is to care for expansion due to heat.

Valve stem guide: A bushing or hole in which the valve stem is placed which allows lateral motion only.

Valve timing: Proper setting of valve actions with reference to piston position.

Valve train: The valves and valve operating mechanism for the control of fluid flow to and from a piston cylinder machine.

Vane: 1. A weathercock for indicating the direction of the wind. 2. A flat or curved surface exposed to a flow of fluid so as to be forced to more or to rotated about an axis, to rechannel the flow, or to act as the impeller; for example is a steam turbine propeller fan.

Vane type supercharger: A positive-displacement rotary blower having an eccentrically located rotor provided with one or more vanes.

Vapor: A gas or any substance in the gaseous state, as distinguished from the liquid or solid state.

Vaporization: A change of state from liquid to vapor by evaporation or boiling; a general term including both evaporation and boiling.

Vaporization line: separates the liquid and vapour regions on the phase diagram.

Vaporizer: A device for transforming or helping to transform a liquid into a vapour; often includes the application of heat.

Vapour: A gas at a temperature below the critical temperature so that it can be liquefied by compression, without lowering the temperature.

Vapour compression cycle: A refrigeration cycle in which refrigerant is circulated through a machine which allows for successive boiling (or vaporization) of liquid refrigerant as it passes through an expansion valve, thereby producing a cooling effect in its surroundings, followed by compression of vapor to liquid.

Vapour lock: A condition wherein the fuel system forming bubble which retard to stop the flow of fuel to the carburetor.

Vapour pressure: The partial pressure of water vapour in atmospheric air.

Vee (V) radiator: A type of radiator made in two sections joined at the middle at an angle less than 180 deg.

Vent: An opening through which a gas can leave an enclosed chamber.

Ventilation: The circulating of fresh air through any space to replace impure air.

Ventilator: A device for providing fresh air to a room or other space by introducing outside air, or exhausting foul air.

Venting: The discharge of hydrogen from a fuel storage system to the atmosphere.

Venturi: Two tapering streamlined tubes joined at their small ends so as to reduce the internal diameter.

Vernier: A small movable auxiliary scale for obtaining fractional parts of the subdivisions of a fixed scale. The complete instrument.

Vernier depth gauge: A rod type gauge fitted with a vernier and used for checking narrow recessed portions, and shoulders or steps of a machine part.

Vertical band saw: A band saw whose blade operates in the vertical plane; ideal for contour cutting.

Vertical clearance or vertical suction (MB plough)**:** It is the maximum clearance under the landside and horizontal surface when plough is resting on the horizontal surface in the working position. It helps the plough to penetrate in to the soil to a proper depth. *See* Fig.1V.

Vertical conveyor: A material-handling machine designed to move or transport bulk materials or packages upward or downward.

Vertical conveyer reaper (tractor mounted): It is front mounted at the tractor, operated by PTO, raised and lowered by hydraulic control. It is mostly used for harvesting of paddy and wheat crop. Power tiller operated, front mounted vertical conveyor reapers for harvesting of paddy and wheat are also available and used extensively.

Vertical firing: The discharge of fuel and air perpendicular to the burner in a furnace.

Vertical lathe: A type of vertical boring mill, which carries a side head.

Vibrating bell: An electric device having a clapper or hammer, which strikes a bell rapidly when an electric current flows through it. It operates on the principle of electromagnetic attraction.

Vibrating feeder: A feeder for bulk materials (pulverized or granulated solids), which are moved by the vibration of slightly slanted, flat vibrating surface.

Vibrating pebble mill: A size reduction device in which feed is ground by the action of vibrating, moving pebbles.

Vibrating screen: A sizing screen, which is vibrated by solenoid or magnetostriction, or mechanically by eccentrics or unbalanced spinning weights.

Vibration damper: A device to reduce the torsional or twisting vibration, which occurs along the length of the crankshaft used in multi-cylinder engines; also known as a harmonic balancer.

Vibration damping: The processes and techniques used for converting the mechanical vibration energy of solids into heat energy.

Vibration drilling: Drilling in which a frequency of vibration in the range of 100 to 20,000 Hz is used to fracture rock.

Vibration machine: A device for subjecting a system to controlled and reproducible mechanical vibration. Also known as shake table.

Vibration separation: Classification or separation of grains of solids in which separation through a screen is expedited by vibration or oscillatory movement of the screening medium.

Vibration suppression: The prevention of undesirable vibration, either through passive means such as damping or through active techniques involving feedback control.

Viscosity: The resistance to flow of an oil. A measure of the resistance to flow of a liquid. The lower the viscosity, the thinner the fluid is and the more easily it will flow. This is affected by temperature. At low temperatures, viscosity is high, while at high temperatures viscosity is low.

Viscous: Thick; tending to resist flowing.

Viscous damping: A method of converting mechanical vibration energy of a body into heat energy, in which a piston is attached to the body and is arranged to move through liquid or air in a cylinder or bellows that is attached to a support.

Vise clamps: Are made of brass or copper, and used over the faces of the steel jaw to prevent bruising the work.

Visibility: The greatest distance at which conspicuous objects can be seen and identified.

Voids: Vacant spaces such as occur between broken particles of a substance, as stone, coal, etc.

Voltage (v): The force that causes electrons to flow in a conductor. One Volt equals one Joule of work per Coulomb of electrons. The difference in electrical pressure (or potential) between two points in a circuit.

Voltage control: Regulation of the generator output by controlling the pressure (voltage) developed by the generator.

Voltage divider: Usually a resistor that can be tapped or adjusted to provide variations of voltages between its ends.

Voltage drop: The voltage consumed in forcing a current through a certain resistance or group of resistances connected in a single circuit.

Voltmeter: An electric meter for measuring the voltage or electrical pressure of an electric device, such as a battery or alternator, or for measuring the voltage between two points in an electric circuit.

Volume (v): An area defined by measurement of length, width, and height, and expressed in cubic units such as the cubic feet, cubic centimeter, etc. Volume is constant, being a measurement of space rather than a condition of air or gas.

Volume flow rate: is the volume of the fluid flowing through a cross section per unit time.

Volume fraction: The ratio of the component volume to the mixture volume. For an ideal-gas mixture, the mole fraction, the pressure fraction, and the volume fraction are identical.

— — —

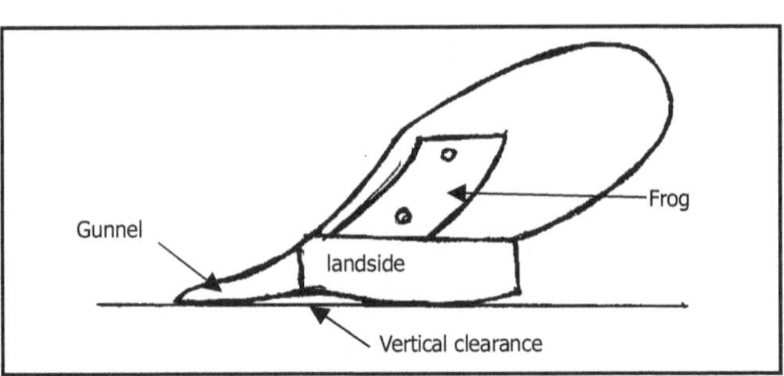

Fig. 1V Vertical clearance

Walking machine: A machine designed to carry its operator over various types of terrain; the operator sits on a platform carried on four mechanical legs, and movement of his arms control the front leg of the machine while movement of his legs control the rear legs of the machine.

Wane: Wane is a defect in a timber or plank.

Washer: A small, flat, perforated disk, used to secure the tightness of a joint, screw, etc.

Washer cutter: A device having a fixed center and either one or two adjustable cutting points for cutting washers from leather, rubber, etc.

Waste heat: Energy, that must be dissipated to the atmosphere from a process such as the heat transferred from condensing steam in the condenser of a steam power plant.

Waste spark system: An ignition system, that energizes the spark each time the piston is at top dead center whether or not the piston is on the compression stroke or on its exhaust stroke. For hydrogen engines, the waste sparks are a source of pre-ignition.

Water (H_2O): 1.A colorless, transparent, odorless, tasteless liquid byproduct of the combining of hydrogen and oxygen, 2. The liquid form of steam and ice. Fresh water at atmospheric pressure is used as a standard for describing the relative density of liquids, the standard for liquid capacity, and the standard for fluid flow. 3. The melting and boiling points of water are the basis for the Celsius temperature system. 4. Water is the only substance that expands on freezing as well as by heating, and has a maximum density at 4°C.

Water heater: A tank for heating and storing hot water for domestic use.

Water injection: A technique for thermally diluting a fuel mixture by injecting water into the hydrogen fuel stream.

Water jacket: The outer casing of an engine block and head, so constructed as to permit a circulation of water between it and the cylinder walls, for the purpose of carrying off heat from the engine.

Water pump: A pump by means of which a forced circulation of the water in the cooling system is produced. These pumps are usually placed at the front of the cylinder lock where they are operated in connection with the fan drive or generator drive.

Water putty: A powder which, when mixed with water, makes an excellent filler for cracks, nail holes, etc. Not suitable for glazing.

Waterwheel: A vertical wheel on a horizontal shaft that is made to revolve by the action or weight of water.

Watt: An electrical unit of power; equals volts multiplied by amperes.

Watt second: A unit of measure of electrical work equal to the rate of one watt expended for one second of time.

Watt-hour: A unit of measure of electrical work, equal to one watt expended for one hour.

Wattmeter: An instrument for measuring electric power in watts; the unit of electrical energy, volts times amperes, combining therefore the functions of a voltmeter and an ammeter.

Wavelength: 1. The length in meters of one complete sine wave of an altering current. 2. In radio, the distance in meters between the maximum points of two successive magnetic waves sent out by a transmitter.

Web of drill: The thickness of a drill at the bottom of the flutes.

Wedge: A piece of wood, or metal, V shaped in longitudinal section, used for producing strong pressure, or for splitting a substance apart.

Weight: The gravitational force applied to a body, and its magnitude is determined from Newton's second law.

Weir: A dam or barrier by means of which the water of a stream is hold back in order to provide a sufficient head of water for power purposes.

Weld period: The time required to go through one complete cycle of a welding operation.

Welding: Uniting of pieces of iron or steel by fusion accomplished by the oxyacetylene, electric, or hammering (forging) process.

Welding rod: Usually 60 cm. long and 3.8, 45 or 2.5 cm. in diameter; used for flowing into the joint to be blowpipe welded. Welding rods are of different composition according to the class of work on which they are to be used.

Welding transformer: A step-down transformer used to produce sufficient instantaneous current to fuse the metals, in contact, through which it is flowing.

Well drill: A drill, usually a churn drill, used to drill water wells.

Wet clutch: A friction clutch, which operates with the application (in a bath) of a lubricant.

Wet mill: 1. A grinder in which the solid material to be ground is mixed with liquid. 2. A mill in which the grinding energy is developed by a fast flowing liquid stream; for example, a jet pulverizer.

Wet well: A chamber which is used for collecting liquid, and to which the suction pipe of a pump is attached.

Wet-bulb temperature: The temperature measured by using a thermometer whose bulb is covered with a cotton wick saturated with water and blowing air over the wick.

Wheatstone's bridge: A method of measuring resistance by the proportion existing between the resistance of three known adjustable resistances and the one to be found, all forming arms of bridge.

Wheel base: The distance between the centre of the front hub and the centre of the rear hub of the automobile.

Wheel bolt: The bolts by which the wheels are attached to the wheel hubs.

Wheel cylinder: The hydraulic cylinder mounted on the backing plate of any brake drum. Pistons in wheel cylinder are forced outward to move the two brake shoes, thereby setting them to the brake drum surface.

Wheel hoe: *See* Fig.1W.

Wheel hub: That part of the wheel assembly containing the bearing cup or races and used for wheel mounting. In front wheels it is mounted to the steering-knuckle spindle. In rear wheels it is mounted on the rear-axle shaft or axle housing.

Wheel puller: A device used for pulling automobile wheels from the axles.

Wheel tramp: Up and down motion of the wheel-mounting spindle, caused by unbalanced wheels or by eccentric wheels and tyres.

Wheel truing: Any operation on any part of grinding wheel to balance it or to change its shape so as to improve its grinding or cutting qualities.

Wheel wobble: The condition existing when a wheel is bent in one place of its original form, an erratic movement of the wheels turning on its spindle.

Whitworth thread: The standard British thread, having rounded tops and bottoms and an included angle of 55 degrees.

Winch: A machine having a drum on which to coil a rope, cable, or chain for hauling, pulling, or hoisting.

Wind indicator: A device that indicates the direction and velocity of the surface wind.

Wind load: The load on a structure due to wind pressure.

Wind tunnel: An apparatus producing an artificial wind or air steam, in which objects (straightner, manometer etc.) are placed for investigating the air flow about them and the aerodynamic forces exerted on them.

Wing nut: A form of nut, which is tightened or loosened by two thin flat wings extending from opposite sides. It is also called as a thumbnut.

Wiped joint: A lead joint in which molten solder is poured upon the two pieces to be jointed until they are of the right temperature. The joint is then wiped up by hand with moleskin or cloth pad while the solder is in a plastic condition.

Wire cloth: A fabric made of wire. The size of mesh is made to suit the purpose for which it is to be used.

Wire gauge: A notched plate having a series of gauged slots, numbered according to the sizes of the wire and sheet metal manufactured; used for measuring the diameter of wire.

Wire glass: A type of window glass in which is imbedded wire of coarse mesh, to prevent scattering of fragments should the glass be broken.

Wire nail: A nail made of wire and having a circular cross section.

Wire saw: A machine employing one or three strand wire cable, upto 16,000 feet (4900 meters) long, running over a pulley as a belt used in quarries to cut rock by abrasion.

Woodruff key: A semicircular or semi elliptical key flattened on the sides, for use in a keyway cut by bringing a rotary cutter against the material.

Work: The transference of energy that occurs when a force is applied to a body that is moving in such a way that the force has component in direction of the body's motion, it is equal to force times distance.

Working drawing: A drawing, which contains all dimensions and instructions necessary for successfully carrying a job to completion.

Working edge: In planning a piece of wood, one of the wide faces is first trued and called the 'working face' , then an edge is trued square with the working face and is called the 'working edge'.

Working fluid: The fluid to and from which heat and work is transferred while undergoing a cycle in heat engines and other cyclic devices.

Working load: The ordinary load to which a structure is subjected; not necessarily the maximum load, but the average or mean load.

Working pressure: The pressure at which the equipment is designed to function.

Working unit stress: The ultimate stress divided by the factor of safety.

Worm gear: A set of gears which has the pinion or worm shaft set at right-angles to the ring or driven gear shaft; usually the worm is made in hour-glass form (larger at the ends). It may contact either the upper or lower surface of the ring (driven) gear.

Worm and gear steering: Consists of a worm on the lower end of the steering gear shaft meshing with a worm gear on the cross shaft.

Worm pitch: The number of worm threads per inch.

Worm shaft: The steering gear shaft, on one end of which is the steering wheel while the other end carries the worm gear.

Worm threads: These threads are of the acme type, having an included angle of 29 deg., but are usually made deeper than the standard acme thread.

Wrench: Common types are adjustable wrenches, monkey wrenches, double-end S wrenches, and socket wrenches. A tool exerting a nut or bolt.

Wrist pin: The journal for the bearing in the small end of an engine connecting rod which also passes through piston walls, also known as a piston pin.

— — —

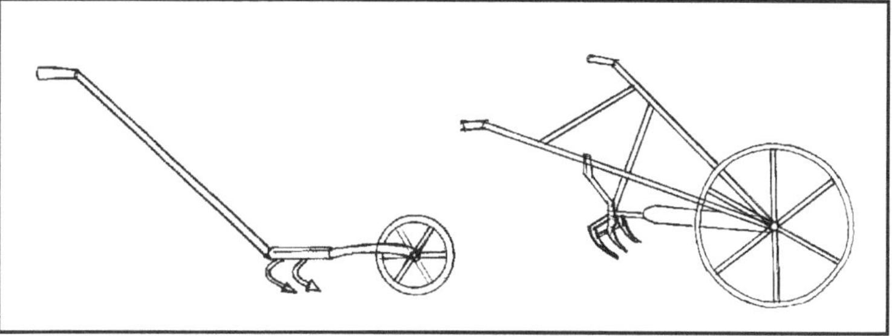

Fig. 1W Wheel hoe

X engine: An in line engine with the cylinder block so arranged around the crankshaft that they resemble the letter X when the engine is viewed from the end.

Yield point: Unit stress at which the specimen begins to stretch without increase in the load.

Yield strength: The stress at which a material exhibits a specified deviation from proportionality of stress and strain.

Yield stress: The lowest stress at which extension of the tensile test piece increases without increase in load.

Yield: That stress at which material exhibits a specified deviation from proportionality of stresses and strain.

Young's modulus: The ratio of a simple tension stress applied to a material to the resulting strain parallel to the tension.

Zero bevel gear: A special form of bevel gear having curved teeth with a zero degree spiral angle.

Zero: The numeral, a cipher, the lowest point.

Zeroth law of thermodynamics: States that if two bodies are in thermal equilibrium with a third body, they are also in thermal equilibrium with each other. By replacing the third body with a thermometer, the Zeroth law can be restated as two bodies are in thermal equilibrium if both have the same temperature reading even if they are not in contact.

ZEV: Zero Emissions Vehicle

Zigzag rule: Derives its name from the manner of opening and closing. Often referred to as a folding rule. Made up in 6 in, sections total from 2 to 8 ft.

Zinc: A bluish white metal principally used kin galvanizing and in a making alloys. Melting point 419°C and boiling points is 40°C. It is not very ductile at room temperature but can be drawn into wires or rolled into thin between 110 to 150°C at about 200°C it becomes very brittle.

— — —

Bibliography

Indian Standard Institution, 'Handbook of Agricultural Machinery Terminology'.

Jain S.C. and Grace Philip, 'Farm Machinery - an approach', Standard Publishers Distributors, Delhi, 2002.

Liljedahl, John.B., Paul K. Turnquist, David W. Smith and Mokoto Hoki. 'Tractors and their power units' fourth edition, CBS Publishers and Distributors, New Delhi, 2002.

Michel A.M. and T.P.Ojha. 'Principles of Agricultural Engineering', Vol.I, Jain Brothers, New Delhi, 1993.

Nakra C.P. 'Farm Machines and Equipment' Dhanpat Rai and Sons.1980.

Rodichev V. and G. Rodicheva. 'Tractors and Automobiles' Mir Publishers, Moscow, 1982.

Sahay J. 'Elements of Agricultural Engineering', Standard Publishers Distriburors, Delhi, 1992.

Thokal R.T., D.M.Mahale, A.G. Powar. 'Glossary -Irrigation, Drainage, Hydrology and Watershed Management' Mittal Publishers, New Delhi. 2004.

Vaishwanar R.S. 'Dictionary of Mechanical Engineering', Jain Brothers. New Delhi, 2001.

— — —

Zeitfracht Medien GmbH
Ferdinand-Jühlke-Straße 7
99095 Erfurt, Deutschland
produktsicherheit@kolibri360.de